8-205-31

Fritz Eggimann

Home: Bachstr. 36
CH-5300 Turgi

A SWITCH IN TIME

A SWITCH IN TIME

AN ENGINEER'S TALE

By JOHN MEURLING
with the aid, counsel, comfort
and collaboration of
RICHARD JEANS

telephony T.M.
Publishing Corporation
55 East Jackson Blvd.
Chicago, Illinois 60604 USA

First Edition: January, 1985

Library of Congress Catalog Card Number: 84-51993
ISBN 0-917845-02-1

Printed in the United States of America

This book is dedicated to all those people at Ericsson, Ellemtel and the Swedish Telecommunications Administration who were and are part of the AXE programme in all its phases.

My special thanks go to my secretary of many years, Sif, who has read and typed the manuscript, corrected spelling, and fearlessly made suggestions for improvements. Sif is also pretty. Which is more than can be said of old Jeans.

I got to know Richard Jeans, the human word processor, several years ago. In preparation for producing a series of telephone network case studies around the AXE system I had laboured hard and long in working out a really brilliant theme. Richard was called in as a consultant and producer and his first words to me in looking over my suggestions (very rapidly I may add) were: "Well, yes, ah, well obviously this will need a lot of improvement." Which is definitely not a socially acceptable way of starting a happy professional relationship.

However, I was soon able to introduce an equally insulting observation, and that is how it has continued. And I and other people at L M Ericsson, found that Richard, though not a telecommunications man, very quickly learned to understand Ericsson (which is more than many of us ever do).

Apart from rewriting everything I do and contributing what social, philosophical and literary values this story holds, Richard is an amateur student of architecture, or houses as he puts it, and a cook. Which is why Janet, my wife, allows him in the house for our writing sessions.

My thanks go to Janet, who married an engineering student. Little did she know what she let herself into. She also encouraged
this book.

J.M.

INTRODUCTION

SWEDEN is terribly Nordic. The winters are very cold, with very short, very dark days. And the Swedes are unnatural. When other people would pull their houses up over their heads and go to sleep for three months, the Swedes have extended bursts of innovative energy. "Ja," they say brightly in English, "here is now the snow. Now, Richard, it's time to work good!"

Happily, John Meurling is only half Swedish, and at least he gets tired some time before midnight on winter evenings. He also has what used to be English standards of hospitality.

And so it was that on a winter evening in 1982, John and Janet and I were sitting in relaxed contentment in front of a dying log fire in their house outside Stockholm, on the edge of primeval, elk-infested, Swedish forest.

Well, I was relaxed and contented, anyway. My advertising agency had been working for Ericsson for two or three years, and I'd been doing much of the work. I had not found it easy.

An old-fashioned public-school education in Wales, followed by classics at Oxford, had done little to fit me for the technological revolution. I passed School Certificate maths, but with such agony that I vowed to call a rectangle an oblong for the rest of my life.

And yet . . . my father began life as an engineer. My brother is a very successful engineer. I had been brought up on Kipling. We had lived in remote and rural locations without much money, where the practical applications of primitive technology were essential for survival. (When you have no electricity, no piped water, no gas, no money, and no help, you become amazingly ingenious. And that's all engineers are—systematically ingenious.) And though in the '50s, the army and the need for a job had drawn me to the well-endowed Southeast of England, I'd never really come to terms with the world of supermarkets and cosmetics, fashion, and conspicuous consumption.

I began work in an engineering company, and eventually helped to found an industrial advertising agency in 1969. Ten years later, I had a working knowledge of the technologies of construction, metal manufacture, printing, electronics, data-processing, fork-lift trucks, plastics, elastomers and agricultural chemicals.

I also had (and have) astonishing gaps. I can't drive, for example, and I can't always get the tops off jam jars.

Telecommunications was the most difficult technology I'd ever tried to get to grips with. Digital switching, pulse-code modulation, time-division multiplexing, opto-electronics—I found (and find) these, and the other elements of modern telecommunications, difficult.

I forget what I'd been doing, that day in 1982, but I'm sure it was arduous, a day spent bridging language and experience gaps—mine on technology, Ericsson's on advertising. It wasn't a day spent with John, who merely picked me up at the end of it, and drove me off into the darkness to have dinner at home.

John's days are also pretty arduous, and by the evening we were both winding down. Idly, I tried to put AXE into context for myself, and asked John how it began, and what his own involvement with it had been.

What he told me, with a few added facts and figures, is what's in this book. It fascinated me then, and it fascinates me now, to be connected, however peripherally, with one of the unquestionable and unquestioned turning points in the way the world is managed. For that is what AXE represents, a significant coup in the information revolution which is transforming the world at incredible speed. If Ericsson had not brought the nervous, tricky business of digital telephony into top form, somebody else would have; so much is clear even in this book. But it *was* Ericsson which did it, and AXE is *still* setting the pace.

But, at the time, we had no idea of writing a book.

Months later, I'd just read Kidder's astonishing bestseller, *The Soul of a New Machine,* when I met John in London on his way from Stockholm to the States. He was tense and distracted, over-committed to what he was doing in Ericsson at the time, and clearly in need of a counter-irritant.

Absolutely with malice aforethought, and hoping he'd get the point, I gave him *The Soul of a New Machine* to read on the aeroplane.

He got the point.

When next I saw him, he was determined to write a book on the development of AXE. He hoped for a four-month sabbatical to do it.

When next I heard from him I was on a week's holiday in Wales. He, I think, was also on holiday, in the South of Sweden. A call from my office asked me to ring him there, and eventually I spoke to him.

He could not, he said, be given a sabbatical. He would not, he said, give up the idea of the book. But since it would have to be a spare-time occupation, he would need a collaborator. Would I like to be that collaborator?

I would. And that's what I'm doing in these pages. The

project has been impossible from the start. It has been written furtively, in snatched weekends, or in occasional hours stolen from legitimate business. But it's been hypnotic.

Maybe it will never be published. If not, it's a pity—not because of its literary merits but because of its subjects: AXE; the life of an engineer; and the marketing of technology. Any hope for humanity lies in the wise selection of worthwhile goals, and the strenuous application of technology to achieve them—and this is the story of a triumph of technology in achieving a benign goal.

We've written the book together, but it's John's story, and he's telling it. Any personal comment I've wanted to make is in brackets and identified. A lot of people have helped me, and my thanks go alphabetically to Conrad Bird, Sandra Cronin, Kim Fawkes, and Gary Howells.

If cheerfulness will keep breaking in—well, we've tried to keep the telling light-hearted. It's a mistake to be too obviously serious about important things.

Richard Jeans
London, October 1984

EXPLANATION

THIS story is about a system. A system can be different things to different people and in different applications. In this story, the system is a complex, large and expensive set of machinery used to build telephone exchanges. It is called AXE. AXE is a switching system.

The first reason for writing this story is that the development of this piece of machinery and its introduction into the world markets represent a major exercise in engineering. It is not unique for its size, or for the degree of success it has reached. It is not unique for the methods of direction and management style that were employed, or for the speed with which it was completed. It is not unique in the attention it has gained in the telecommunications community. But it does contain elements of all those characteristics and I believe the combination is unique. A strong reason for writing it is also, of course, that I was actively engaged in different parts of the programme, from the very beginning and over a period of about ten years, and—I wanted to write it. I believe it is worth telling also as a story of engineers meeting a great challenge, their response to this challenge, and the hard work and satisfaction it provided.

But it is not the official story of AXE, it is very much my story. So it is not a management report, but rather the story of one man's experience, at a vaguely-defined middle-management level, as one of a large team. And in being this it is also one engineer's tale of being an engineer. It is largely set in the engineer's world—the engineer in modern industry—and describes something of his life working with system design, with production and with marketing.

The setting is L M Ericsson, a telecommunications company in Sweden, and my employers for over 30 years. L M Ericsson is the parent company of the Ericsson Group, which includes some 150 companies all over the world. It is a truly international company, with more than 80 percent of its sales outside Sweden—and this side of its business is naturally very much reflected in the work and environment of its engineers. The AXE system is the latest in a long line of products designed for a particular international market—the telecommunications Administrations and telecom operating companies around the world. We call this business area "public telecommunications". In this area, it is a tradition at Ericsson, and indeed a necessity, that marketing and sales are to a very large extent handled by engineers. Our customers are large, professional organisations, in which the decision-making rests on evaluation and studies performed by technically trained staff—so it is natural that the provider of products and services should meet them on common ground and equal terms. The book deals to a large part with marketing—with engineers going to market with their new product.

To those of us who were, and the many who still are, actively involved, the AXE development programme is a thrilling story. It has been full of excitement, unforeseen problems, and one or two setbacks. Part of the excitement and, of course, an ever-present factor in our lives, is provided by the competition, and a fair amount of space is devoted to our competitors in the field and to the many competitive situations in which we have been involved.

Although the story concerns a system and although much of the narrative deals with technical matters, this is not a technical textbook, and, we have tried to keep the technical part down to a minimum. When it has been necessary to bring in technical matters we have tried to treat them in easily understood terms, hoping that non-engineers will also find the book readable.

One more thing. Quite often in this book I have associated myself with Ericsson, and said that "We" did this or "We" thought that. In fact, all the judgements in this book are mine; received opinion in Ericsson may be different.

So this is my story about AXE and some of the people who took part in this exciting episode. Several other people have helped with background and have taken time to recall events that might have been forgotten—and some that maybe should have been forgotten. But basically, it is my story, and any opinions expressed are my own. Maybe some day the official story of AXE will be written—it is certainly worth it. It will no doubt be longer—and probably duller.

John Meurling
Stockholm, October 1984

CONTENTS

Dedication V

Introduction VII

Explanation XI

CHAPTER 1. How it all began 1

Earning (a bit) and learning (a lot)

A first lesson from the gospel according to St. John

The company of Lars Magnus Ericsson

The years of decision

The roots of change

CHAPTER 2. Birth of a world-beater 38

The requirement specification

The decision

Getting the message

A second lesson from the gospel according to St. John

Here endeth the second lesson

Decisions, decisions

It works

CHAPTER 3. Making it in the market place 89

Launching the new system

Marketing and the eternal truths

Las Marimbas

CHAPTER 4. Backswing and follow-through **134**

Into the world markets

Jumping on the digital bandwagon

The end of the first ten years

CHAPTER 5. Where do we go from here? **155**

Growing up

Showing off

Telerica

CHAPTER 6. As far as we can see **166**

The competition today

Final chapter—but not the end

CHAPTER 1: HOW IT ALL BEGAN

EARNING (A BIT) AND LEARNING (A LOT)

IN 1950, I was a student at Chalmers Institute of Technology, aiming for a degree in electrical engineering. Life was quite busy with various student activities, and I had got married that summer. My wife, my parents, and the faculty were all exerting pressure, and had made it abundantly clear that I was expected to get my degree by the summer of 1951. I was not so sure—academic studies were not all that fascinating, and looking through my report book it was clear that the grades I had made were pretty flat. After a few fours and fives during the first two years, it was all threes, or passed. And there were still a lot of blanks, showing that I had yet to pass a number of examinations. Still, life went on, and I enjoyed being a student. At that time, the class arranged the traditional tour of study of Swedish industries—we spent a week visiting major companies to get to know them, as a help in making up our pure little minds about where to look for a start in engineering, and where to pass through the gates into the hard, cold, commercial world of industry. Not all of us would, of course, be going into industry, but for many of us it was to be the first step in our careers as professional engineers.

Among the companies we visited was L M Ericsson in Stockholm. I don't remember very much of what happened (there had been a bit of a party the night before) except that we were told of the great opportunities there, and we were given lunch. Though the lunch was excellent, I didn't come away from that visit with my mind made up or with a great urge to go into telephony.

By the following summer, 1951, I had, by hard work and the help of lenient and understanding professors, and God, passed (threes) all the required examinations. All that remained was to complete my thesis. During the previous two years, I had been pursuing a course in nuclear physics. Much of it was well above me, but it was a new science and I had some vague idea that it might eventually provide an interesting career. I had also chosen nuclear physics as an area for my thesis, and been given by my professor a project involving the design of an instrument to measure a specific phenomenon in gamma radiation. I spent the summer doing this. By August, the apparatus was built and I had written most of the paper. All that remained was to test and calibrate, and to do this I needed a specimen of active material—gold. I had ordered a piece of the proper stuff from a laboratory in Germany, but they had not delivered the goods, and I couldn't get a firm answer on when to expect it.

So I had a bit of a problem. I was hanging around waiting for the gold, when what I wanted to do was get a job. The idea of going to work, with all it entailed in regular hours, being at least moderately presentable in dress and sobriety during long hours every day, and having boss persons telling me what to do, was not all that attractive, but I needed the money. My wife and I couldn't live even half decently on her pay, which was a pittance anyway since she, too, had just started out on her career as a lowly librarian.

Four years of academic studies and long hours in the laboratory had taught me something about myself—I had come to realize that I was not the type to go into theoretical work, and I did not have the patience for the laboratory. I thought I must look for something in engineering that was closer to producing tangible results.

In 1951 we did not have television in Sweden. Today, many of us feel it should have stayed that way, but it didn't, and planning to build a system had started. Since it was a new field, I decided they needed my talent to get the thing right, so I wrote to the Swedish Telecommunications Administration, explaining my situation, and offering my services. They were surprisingly slow in responding, but eventually I had a letter back. I was not offered a job, but I was invited to come for an interview, at my convenience. The next day I travelled up to Stockholm and had a very interesting talk with the senior technical manager of the Television for Sweden plans. The job I was offered to start on concerned field work. In order to site the future television transmission towers, a large programme of transmission field effect measurements had been planned —and I would be welcome to join this pioneering team and add my expertise and enthusiasm (both characteristics as yet undocumented) to ensure that my people got the best television system in the world. As far as I can recall, this was on a Wednesday or Thursday in the end of August. When I suggested I might start working the following week, the blow fell. They had no position available now. The plans had been approved by the government, and were scheduled to start in January—next year. Thus ended my first and only attempt at working my way into the Swedish state system.

Back on the street, I did some quick thinking and put in a call to my father. He had worked as an accountant in an electrical manufacturing company for years, and apart from being a wise old man (he was younger then than I am now) he knew a lot of people in industry. His advice was that I should go out to L M Ericsson and see a colleague of his, through whom I could get to the part of the company which might be interested.

At the end of that visit, I had (a) had telephony explained to me; (b) been told that an engineering degree from university really didn't carry much weight at L M Ericsson, since telephony was a specific and difficult art that few could master; and (c) a job, at a ridiculously low salary. I had also met my future boss and agreed to start working the following week, if I had my degree. For even having been told that the title of "civil engineer", as we are called in Sweden, was not worth much at my future place of employment, after four years of sweat and tears I really wanted that piece of paper.

So I still had a problem, and back in Gothenburg the next day I had a heart-to-heart talk with my professor, explaining what I had done and my dire economic straits. I also pointed out that in my opinion the actual testing of the machine I had designed was just a detail after all the care and research I had put into it. At the end of that conversation I had become an engineer. Lovely man, the professor. Though I don't for a minute think he really shared my conviction that the machine would work, he performed a truly Christian act. Some years later I heard from a friend who had gone to work at the physics department that he had seen my apparatus stored away in the loft gathering dust. Thus ended my venture into nuclear physics. As I said, I never really understood much of it anyway.

The following week I travelled to Stockholm, was offered temporary lodging with my grandmother, and went to work for L M Ericsson as a system designer of crossbar exchanges. I understood virtually nothing of what it was all about, but I had a job.

The rest of the family, Janet and some furniture, moved to Stockholm about a month later when we had found a flat. Janet also got a job at L M Ericsson, in the library. For me, it was a complete change of existence. Gone was the carefree student life. I was now a member of the vast community of office workers.

I was attached to a group of about eight engineers, working on both the design and the marketing of systems for the Finnish market.

I spent most of the first year unravelling the complexities of telephone switching, and was then lucky enough to get some experience in the field. The Helsinki Telephone Company was one of the first customers for our new crossbar system, and towards the end of 1952 I was sent over to help on some traffic studies we were conducting.

My visit lasted seven months! The system had a couple of irritating faults, and I was told to try to locate and correct them. Eventually, with a lot of help from the Helsinki people, I managed to find out what was wrong and work out how to put it right.

The exercise lasted throughout the winter and spring of 1952/53, and had three results.

First, the Helsinki Administration took to Ericsson and the new crossbar system overnight. They even paid the balance outstanding on the contracts which they had been withholding because of the less-than-perfect performance of the exchanges.

Second, some of the success rubbed off on me. My boss developed the view that I understood switching.

Third, I actually did get some accelerated insight into switching. And working long days (and nights) closely with a customer, I gained a better, more intimate understanding of switching in action than I could ever have acquired at my desk or in the labs with Ericsson at Stockholm.

When I got back, I was involved in some early attempts to design a rural network system, still in close contact with Helsinki. But in the autumn of 1953, I was developing other preoccupations. We were expecting our first child, and we had decided to buy a house.

There was no difficulty in finding a house. Finding the money was another matter. I approached my boss, and asked politely whether the company would consider helping a trusted employee of long standing (well, two years) with a loan. The amount we needed would mean little to the company; to us, however . . . and so on. My boss was understanding, and promised to take the matter up with the proper authorities.

A week later, I was summoned to see one of the company's directors. I was ready for this. I had calculations and a budget plan, and a sob story of our need for green grass and fresh air for the child (when it arrived).

I sat before the great man, looking, I believed, serious and business-like, with my sheet of calculations at the ready.

But at his first question, the paper went back into my pocket.

Would I, he asked, like to move to Mexico? I managed to say that it sounded most interesting, and asked for a day to consult my wife. Within an hour, we had forgotten about buying a house and the child's needs for green grass. We were busy finding Mexico on the map and discussing the best way to learn Spanish.

In August 1954, we left for Mexico with two-months-old Peter.

We stayed there for eight years, had three more children, and eventually bought a house.

I joined Teléfonos de México, the telephone operating company, in the engineering department. At that time, Teléfonos de México was jointly owned by ITT and L M Ericsson, each with 50 percent. Later, in 1958, both companies sold out, and Teléfonos de Mexico became Telmex, and 100 percent Mexican-owned though it was still a private company.

I arrived as the first crossbar equipment was coming into Latin America. During the first two years, I was mainly concerned with testing new exchanges.

On my first day in the office, my new boss took me down to an installation in the same building, introduced me to two Mexican technicians, and explained what was to be done. A switch had been installed to distribute traffic from the long-distance operator's office to the many local exchanges around Mexico City. The equipment was new. It had been installed according to the documentation, but nobody could understand the test procedures. This was partly because the procedures were written in English and Swedish, and partly because they weren't all there.

The job seemed straightforward enough, apart from one problem—communication. I knew about 20 words of Spanish.

It took us a long time to get that switch into service. It became rather involved when we first had to modify the circuits in all the surrounding exchanges, and then the manual boards at which the operators handled the calls.

But I did learn Spanish—by osmosis, and out of necessity. My colleagues, the technicians, were very helpful, and in time we became good friends. We set certain linguistic priorities. First, the cusswords, in which Spanish is very rich. Second, the basic items of food and drink. Third, the tools and the necessary vocabulary to read and discuss the intricacies of circuit diagrams.

It worked. Within a year, and with very little systematic study with books or professional teachers, I was able to keep my end up in everyday conversation, and to write technical memoranda.

Gradually, I became more and more involved with planning, and with requirement studies on new equipment for the Mexican network. We began to plan for automatic long-distance dialling, and eventually we were installing toll exchanges all over the country—first for operator dialling, and later for full subscriber dialling. We also started talks with

AT&T in the US to develop common plans for dialling to and from the US and Europe.

It was interesting, and I enjoyed it. It gave me some unique experience, which I have drawn on for the rest of my life.

I was working in an operating company, a company providing a service to subscribers, *not* a company providing hardware to another company. Knowing how it works from the customer end is invaluable to a manufacturing organisation, and I *was* a customer—of Ericsson and ITT. In Helsinki, I'd found it fascinating to see a telephone network actually working. In Mexico, I was part of it.

A network is very complex. It's alive, it changes and develops continuously as it grows, and as it absorbs new technologies. New pieces of equipment, like switches, are normally grafted onto an existing system, to extend it, to add new trunks for example, or a new handling procedure for long-distance calls. The new technology must fit into an environment which may already incorporate equipment of varying age and design, and the manufacturer of new equipment *must* take all this into account if he's going to get it right and sell it successfully.

My experience in Mexico was practical, hands-on problem-solving, with little actual design or theoretical work. I think this aspect is important, too. I know only too well the feeling of desperation in a situation where you urgently need spare parts or a manufacturer's suggestions on modifying a circuit. While the boys at the manufacturers are getting more and more involved in profound discussions of how to go about it, and are having cups of coffee or their annual vacation, the chaps at the sharp end are jumping up and down, and the operating company is losing patience and revenue.

But by 1961, I was getting restless and ready to move on. Although Janet and I liked Mexico very much, enjoyed the life and had made lots of friends, we had children to educate. It was obvious, too, that Telmex would be forced to reduce the number of foreigners (i.e. Swedes) working there. Most important of all, perhaps, was my own desire to get back close to systems development. The reason was electronics.

Throughout my tour in Mexico, I kept in fairly close touch with the developments going on at L M Ericsson, and also, through reading and meeting people, in the other large manufacturing companies.

Switching was still largely electromechanical, though electronics was coming in for some specific functions. But a whole new concept was emerging—the application of computer technology to the control of switching systems. It was called stored program control, SPC, and I wanted to know more about it, and to be part of it.

We left during the summer of 1962, a large family with lots of luggage. We left a country we had grown to love, and a lot of friends and colleagues we would miss. Sometimes, we still regret the move back—in November, say, when Sweden is so cold, wet and dark.

L M Ericsson had changed while I was away. Marketing, particularly, had acquired increasing importance, and into marketing I went. Systems design was still my background, but I was no longer working actively as a systems designer. I was given responsibility for marketing systems in various territories, the last of which included North America and Oceania. Our most important customers in these areas were Australia, Mexico and Fiji.

In 1970, I transferred, and became a one-man staff department, as assistant to the marketing director of X Division.

To explain what *that* means, it's time to look at the company which had supported me as a wanderer over the face of the earth.

And to make sense of Ericsson, you need to know a little telephony.

A FIRST LESSON FROM THE GOSPEL ACCORDING TO ST JOHN

The blasphemy is not mine. In the early years of the AXE project, it was my job to sell the concept, internally and externally, day and night, at home and abroad. I probably became a bore about it. Certainly I became absolutely identified with it, and with its presentation. My kindly colleagues expected me at every meeting to propound "the gospel according to Saint John". Their words. Their blasphemy.

From A to B: networks in action

This book is not about telephones. It's about telephony—or at any rate part of it.

It's about a particular method of controlling and operating the "traffic" over a telephone network.

In some ways, the traffic in a telephone network is like the traffic in a railway network. Some of the words—for example, lines, switching, and various others—are common to both activities.

With railways, it's easier to see the switching and control going on. Although a train has a driver, he doesn't actually decide where the train is going. It can go only along the railway lines that are laid for it. Its direction is chosen by the control staff, who alter the points on the line to switch the train as it approaches. It's easy to imagine how such a process can be computerised. And, indeed, it's easy to see how a railway network could be described as one vast computer—just as a telephone network can be.

But a telephone network is infinitely more complex and subtle—like a railway network where every traveller has his own personal railway station. The telephone you use daily in your home or your office is the tip of a gigantic iceberg. The telephone itself is just one of 550 million terminals connected to what has been called the world's most complex machine.

This was not always the case. When telephony began, it could only provide local service, you could call the subscribers who were connected to the same exchange as you were. But then these local exchanges became connected with each other, city became connected with city, country with country—until today you dial direct to most countries from most other countries. It is this linking together into a world network that makes the whole thing into one immense machine—a machine with millions of working parts.

Telephones provide the man-machine interface, but in a national network, telephones account for only about 4 percent of the total cost of building the network.

Where does the rest of the money go?

If you move into the network from your terminal—the telephone—the first thing you find is a pair of wires connecting the telephone to a distribution box outside your house in the street. Up a pole, perhaps, or in a pillar on the pavement. This box collects the wires from several phones and runs them through a cable to a distribution cabinet—the boxes you can see on street corners.

All this outer portion of the network is called the secondary network, and in cities it is now normally in underground ducts.

Within the distribution cabinet, the individual pairs of wires (usually just called pairs) are cross-connected to pairs in the cables of the primary network, which are run to the local exchange. These cables, too, normally run in underground ducts, or even tunnels. (Most city streets are more or less hollow!) At the exchange, they terminate on something called the main distribution frame MDF, or main frame—not, of course, in this context a large computer. The individual pair of wires from your home phone has been extended all the way through to the exchange.

The primary and secondary networks together form the subscriber network. The cables and cabinets and other bits and pieces are sometimes called the outside plant, and sometimes the local network, or even the local area network—though these days a local area network, LAN is the term for a device such as a coaxial cable which provides terminals with access to computers, word-processors, printers, and so on, within an office.

Not surprisingly, the exchange is called the local exchange (or in the US, a central office).

Within the exchange, on the main distribution frame, the wires from the primary cables are cross-connected to wires in cables that are run up to the actual switching system and terminate on line circuits. When you lift your telephone handset, the line circuit detects that you want to make a call, and prompts the exchange to connect your pair of wires to a device which sends a tone back to you—the dialling tone.

When you dial, the first few digits are enough to tell the exchange where you are trying to call. If you are dialling somewhere outside your local area, the exchange finds a path through the switches to an outgoing trunk circuit.

From here, it's easy to see how your call is progressed. The trunk circuit is another pair of wires in an underground cable running to another exchange. There, they terminate on another cross-connecting frame, and are connected to an incoming-trunk circuit waiting to receive them. When the connection is made, the originating exchange sends over the last few digits you have dialled to tell the terminating exchange which subscriber line to connect. When this connection is, in turn, established, the exchange rings the called subscriber (your "party"), and when your party answers, the connection is complete and you can speak to each other.

This is a pretty simple example, involving just two exchanges. In a diagram, it looks like Fig. 1 on page 10.

In large cities, the network is much more complex, of course. There may not be direct trunk lines between all the local exchanges. Instead, some calls will be routed through on an intermediate exchange called a tandem exchange (in US, tandem office), and setting up a call involves three switching systems (originating, tandem, terminating).

When you make a long-distance or international call, yet another level of exchange comes into play, the toll, or transit, exchange. You usually have to dial an extra set of digits to

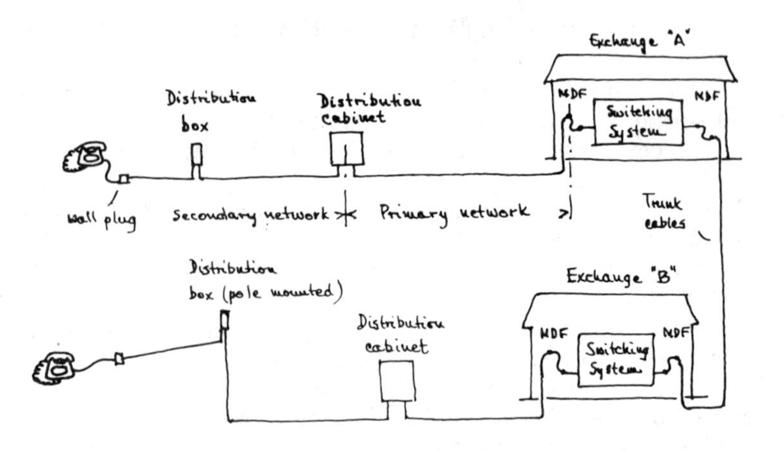

Fig. 1. The basic components of a city network.

start with, and these warn your local exchange to connect you to a transit exchange which takes over the setting up of the call. In the city or country of destination, a corresponding transit exchange receives the call on an incoming inter-city or international trunk and sets up the connection. You see the principle.

So, to come back to the original question, where *does* the money go to? Here's the answer for a typical national network. See Fig. 2 on the following page.

In 1980, the average total investment per subscriber in the traditional sort of network we have described was about 2000 US dollars.

But that really *is* an average—it costs a lot more in rural areas, for example, where outside plant is more expensive per head simply because of the distances involved. In a densely-populated city it may cost less.

Avoiding traffic jams in the network

It may not seem like it when things go wrong—as they inevitably do sometimes in the biggest machine in the world—but making a connection through the network is pretty fast and getting faster.

How do you avoid traffic jams?

The laws of probability, statistical calculations, and

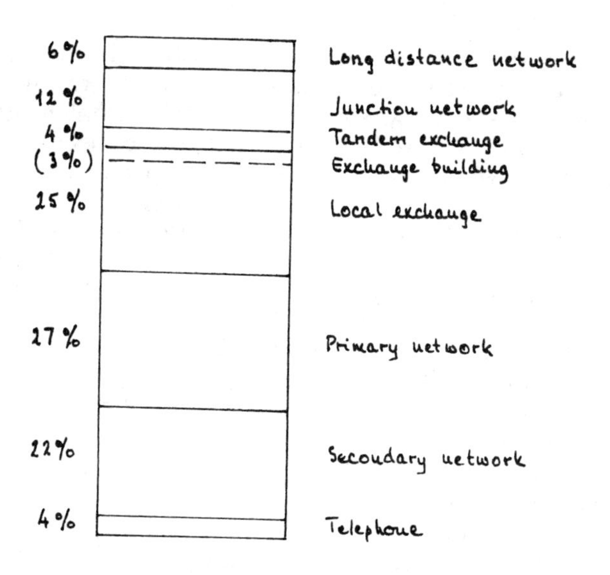

Fig. 2. Typical cost distribution in a telephone network.

straightforward observation of the habits of telephone users form the basis of telephone traffic theory. And telephone traffic theory is the basis of the discipline of traffic engineering.

Say you have a hundred subscribers connected to an exchange. It's very unlikely that they'll get an urge to make a call at the same time, and it's very unlikely that 50 of them are talking to the other 50 at the same time.

Which is just as well, because if telephone exchanges and networks had to be engineered so that every subscriber could *always* without exception put a call through the system, most of the equipment would be idle most of the time. Traffic engineering decides the optimum compromise between cost and convenience.

Traffic engineers work out how much equipment is needed to handle a given amount of traffic, in terms of number of calls and average call duration per subscriber. The figures for various types of subscribers, residential, say, or business subscribers, are known through measurement and experience, and the aim is to provide a "grade of service" good enough to keep subscribers happy. A given amount of equipment will guarantee, for example, that a subscriber can make 998 calls out of 1000 successfully; the failure rate will be less than 0.2% ($p \leq 0.002$).

With that principle in mind, you will see immediately that, whether an exchange is manual or automatic, for a given number of subscribers you need a much smaller number of cord circuits. (A cord is a device an operator uses to complete the circuit which connects you to your party. At its simplest, it's a pair of wires with a jack plug at each end. To connect you, the operator pushes one plug into your jack and the other into the jack leading out to your party. Each cord can, of course, be used in only one complete circuit at a time. Within the exchange, each cord provides one "cord circuit" available to set up calls. An exchange with eight cord circuits can handle eight calls simultaneously.)

An exchange dealing with 100 subscribers may need only 8 cord circuits. But if the exchange gets larger, say 400 lines, 30 cord circuits may be needed, and one operator can probably no longer cope.

How can you tell? And how many more operators?

There are two ways of measuring traffic—in erlangs, and in CCS.

The erlang is named after the Danish traffic mathematician, AK Erlang. It is defined as occupation minutes per hour. (To a traffic engineer, you start to occupy a line when you pick up your phone to dial.) If your telephone is busy for 6 minutes per hour, your traffic is $6/60 = 0.1$ erlangs. If you talk for an hour, you've generated a whole erlang of traffic.

The other standard unit, CCS, measures traffic in conversation lengths of 100 seconds per hour. One CCS means you use your phone for 100 seconds in the hour.

An hour contains 3,600 seconds, so 1 erlang = 36 CCS. Or, if you like 1 CCS = 1/36 erlang.

Now consider your 100-line exchange. If each line is busy for 3 minutes per hour, the traffic on each line is 0.05 erlangs. Multiply that by the number of lines, and the traffic the exchange is handling is 5 erlangs.

From there on, it's simple. To find out how many cords you need, you just consult a traffic table. To handle 5 erlangs, with a grade of service of $p \leq 0.002$, takes 8 cords.

So far, so good.

But it would be quite unsafe to assume that 8 cords need only one operator!

We've described the traffic as 0.05 erlangs per subscriber, meaning that each of them is using his phone for a total of 3 minutes in the hour. But the amount of work the operator has to do obviously depends *not* on conversation time, but on the *number of calls.* If you divided your 100 subscribers into 50 pairs, and each pair had one 3-minute conversation in an hour, there would be 50 calls for the operator to answer, set up and disconnect. But if each pair had three 1-minute calls instead (still 0.05 erlangs!), the operator would have to handle 150 calls in the hour. Life will be so hectic that subscribers will find they have to wait for the operator to answer them.

Such waiting is permissible—up to a point. It, too, is governed by a grade-of-service parameter, normally written as $p\ (t < 1s) = 0.02$. In other words, the probability of waiting more than 1 second for the operator to answer is 2 percent—it will not happen more than twice in every hundred calls.

But, in fact, in everything we've said, "calls" includes not only the calls that get connected and result in conversation, but also calls when the called party is busy, or doesn't answer, or where the phone is answered but the person you want to talk to isn't there. These are "call attempts", and though they may result in little real traffic, they all make work for the operator.

In the end, *the total number of call attempts,* successful and unsuccessful, determines the number of operators—or, in the case of an automatic exchange, the number of devices needed to control the calls.

That still isn't the end of the traffic story.

Both traffic and call attempts have been discussed in "per hour" terms. What hour? Well, certainly not any old hour. We have to engineer equipment, or dimension exchanges, to provide a service to real subscribers—who, for example, spend some hours eating or sleeping. In fact, leaving aside major unforeseen upheavals (earthquakes, city fires, acts of war and Mother's Day) which cause a lot of people to try to make calls simultaneously, telephone traffic follows certain patterns.

These patterns are established by measurement and observation—important functions in all telephone administrations. Traffic varies by time of day, day of the week, and time of year. Business phones generate more traffic than residential phones, and so on.

What we're interested in is the maximum traffic level for which service must be provided—not just any old hour, but the precise "busy hour" of the year in any environment. The variations from environment to environment are enormous, but as a matter of interest, the busiest hour in a large city centre often occurs during late morning, on a Friday, in September.

Of all the hours of the year, the busy hour is the important one, and all further discussions of traffic figures refer to this busy hour. The two most important dimensions of traffic are traffic itself (measured in erlangs or CCS) and busy hour call attempts, BHCA.

They're easiest to understand in small exchanges, but as the AXE story unfolds, we shall change scale. The 100-line example will grow, and we shall be talking of exchanges where 10,000 lines are typical. Traffic for an exchange of this size will be of the order of 1000 erlangs and 40,000 BHCA—ten call attempts per second!

Look, no hands

It's because an exchange has to handle these large amounts of traffic that automatic switching—switching without an operator—has become necessary.

Many people can still remember the days when telephony was manual—the days when making a call was often some sort of hand-to-hand struggle between you and the operator. In those days, each telephone line ended in a jack at the switchboard—a hole with a small lamp, which belonged to you, and had your number written over it. To make a call, you would lift your handset, and turn the handle of your generator.

The operator, as we've mentioned, had a set of pairs of plugs, called cords. To answer your call, she'd select a pair and insert one of the plugs in your jack. She could now speak to you, and ask you what number you wanted. Then she would push the other plug into the jack with the number you asked for written over it, and ring it. When it answered, she would connect you in and herself out, by throwing a key in the cord.

When the conversation was finished, and you hung up, a lamp in the cord circuit would light up, and she would disconnect you by pulling out both plugs.

With an efficient operator, it was a perfectly efficient system. And although most exchanges in public telephone networks are now automatic, manual telephony doesn't belong exclusively to history books. It's still in use in public networks in some rural areas, and offices and hotels and similar organisations still use it in private exchanges. In fact, new and improved manual exchanges are still being developed and sold for private networks, and for military use.

But the change over to automatic telephony began over

eighty years ago, and it's now virtually complete.

The impetus for the change came from the very rapid growth in the number of subscribers, and the consequent increase in the size of exchange.

Think, for a moment, of the operator's physical environment. Here she is in fig. 3.

Fig. 3. Aunt Selma at the switchboard.

She's handling the switchboard in a tiny exchange with 20 subscribers. The subscriber whose telephone number is 02 is calling the one whose number is 13, and the operator is making the connection. One plug of one cord is pushed into the 02 jack, and she is about to push the other plug into the jack of subscriber number 13. The plugs of the other cords are sticking up in front of her waiting for her to use them on other calls.

In this set-up, a single operator might handle up to about 200 jacks, including all the possible combinations of calling and ringing. But what happens when an exchange has 10,000 subscribers?

Simple, you say. Have more operators.

But it *isn't* quite that simple. It's true that you can divide the jacks up among the operators, giving each operator, say, 200 subscribers to look after.

But those calls coming in may be for any one of the other 9,999 subscribers, and each operator would have to get at all 10,000 jacks to make the outgoing connections. It's not simple, yet at the beginning of the century very large manual exchanges were certainly built.

There are two problems to overcome. One is sheer physical access to the jacks. The other is finding the right jack quickly.

In fact, several attempts were made to adapt manual exchanges to allow several operators to handle calls simultaneously. The most successful was the multiple exchange. In a multiple exchange, each operator looks after a switchboard which takes incoming calls from, say, 400 subscribers. There might be six operators and six switchboards, and *all* the outgoing lines would be connected to ringing jacks on each switchboard—the circuits are multiplied. Depending on traffic, this arrangement would allow six operators to handle 2,400 subscribers—but not much more. For larger exchanges, we would need two operators to set up most of the calls.

But there were other drawbacks. There is the famous, and true, story of a Kansas City undertaker whose business was dying because the city telephone operator directed his business calls to a competitor—her husband. Eventually he found out, got the operator fired, and invented the first automatic telephone exchange. His name was Almon B. Strowger, and there is still a lot of Strowger equipment in operation—even in Kansas City, for all I know.

Inevitably, automatic exchanges swept the world. What goes on in them? Let's look again at the lady in the manual exchange. Three operations are going on.

She is picking up incoming calls (line-finding) and asking the caller what number he wants.

She is setting up the connection outwards to the called party (line-selecting).

And she is controlling the whole operation, ready to tell the caller if the line is engaged, or if there's no answer; making a record of the destination and duration, so that the caller can be charged; identifying any line which is faulty; and so on. Finally, when the call ends, she will disconnect.

At Ericsson, we regard line-finding and line-selecting as true switching operations, separate from the control.

We can draw the operations diagrammatically, leaving the control element out. (The quarter circle, with an "arm", is a common symbol for a switch, or selector. For us, they also represent the operator's arms and hands directing the plugs to the jacks.)

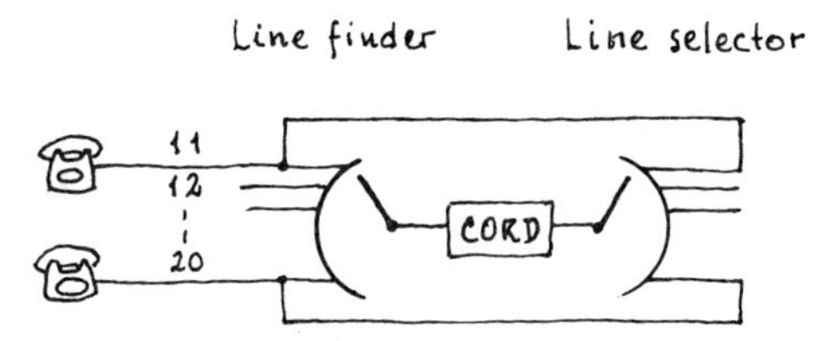

Fig. 4. A 20-line automatic exchange. Note that the wires from the line-selector are joined to those from the line-finder.

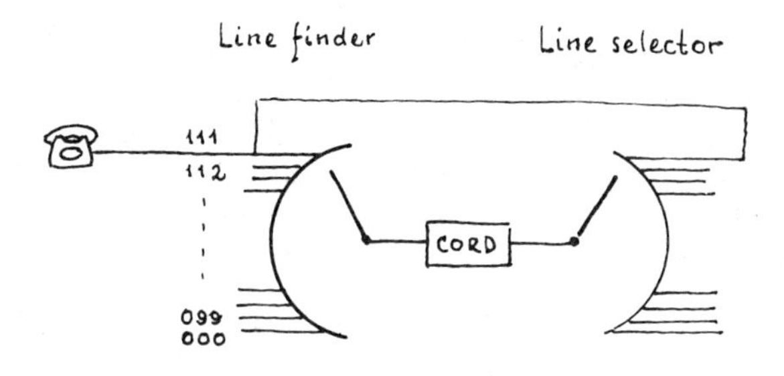

Fig. 5. An exchange for 1000 subscriber lines. Note that there is only one cord circuit so that at a given moment only two subscribers can be conversing.

For an exchange to be able to carry more than one call at the same time, we must build it with more than one pair of selectors and cord circuits. Fig. 6 shows how this is done. If you imagine each selector as some form of flat semicircular device with an arm that passes over a series of contacts along the perimeter, you can see how a number of these selectors are arranged on top of each other. The edge contacts are connected vertically so that each selector can make contact with each of the lines.

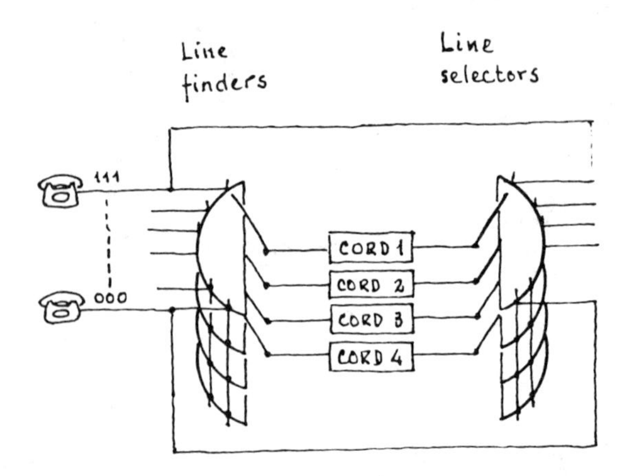

Fig. 6.The 1000-line exchange with four cord circuits, four line-finders and four line-selectors.

The vertical parallel connections are called multiples—the selector outlets are multiplied. The exchange is equipped to handle up to four simultaneous calls.

To simplify matters, and to equip the exchange for a higher traffic carrying capacity, we can reduce the drawing to a diagram in which the number of devices, the cord circuits and corresponding selectors in this case, is indicated. Any telephone engineer, born before 1960 will immediately understand exactly what is meant by the following figure.

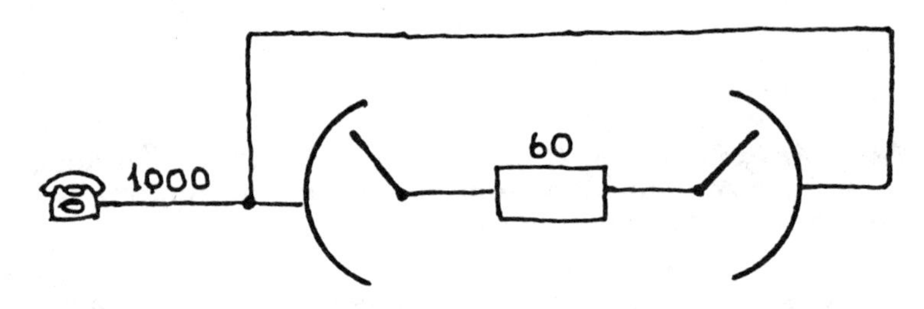

Fig. 7. An automatic exchange for 1000 subscriber lines, equipped with 60 cord circuits.

It's obvious that the selector for the 1000-line capacity is a bit on the large side, like an operator with immensely elongated arms. To avoid the need for such mechanical monsters, the subscribers are divided into groups. Line-finding remains unaffected in principle, but line-selection is in two stages—first the group which contains your party's number, then the number itself.

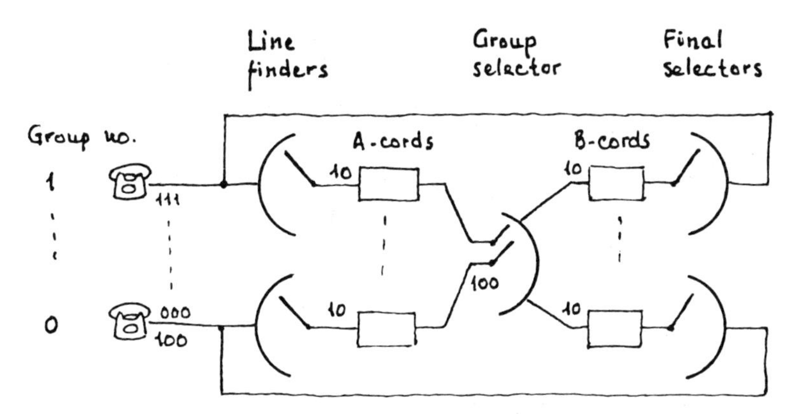

Fig. 8. Exchange for 1000 subscriber lines, divided into 10 groups of 100. Each group has 10 line-finders with A-cords and 10 line-selectors (now re-named final selectors) with B-cords. A group selector has been added, with 100 inlets and 100 outlets.

By incorporating a group selector, we have divided the cords into two categories: A-cords which feed DC to the calling, or A, subscribers; and B-cords which feed the B-subscribers, do the ringing, and so on.

You will see that the first stage is performed by a new switch—the group switch or group selector. This switch is of great importance in the story of AXE. The significant thing is that its operation can be quite separate from line-finding and final line-selection.

The next drawing takes the automatic exchange one step further. See Fig. 9 on page 20.

The line-finders and the final selectors in the last few diagrams have common multiples, or, in other words, each subscriber line appears on a number of line-finders and on a number of final selectors—just as the operators in our original small manual exchange had one common set of jacks in front of her for both answering calls and ringing. Taking first one single group of 100 subscriber lines, we may then draw that group as shown in Fig. 10 on page 20.

What happens up-stream or down-stream of the group selector (line-finding and line-selection) tends to be talked of as happening at the subscriber stage, or *subscriber switch.*

And once the group selector is introduced, dialogue with tandem, trunk or toll exchanges will, of course, be at group selector level. It's the group selector that looks after traffic with the outside world.

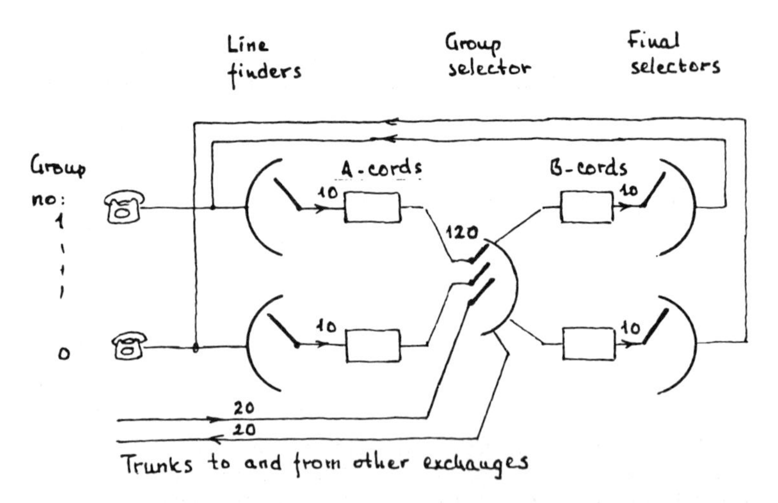

Fig. 9. The same exchange as in the preceding diagram with trunks added to carry traffic to and from other exchanges.

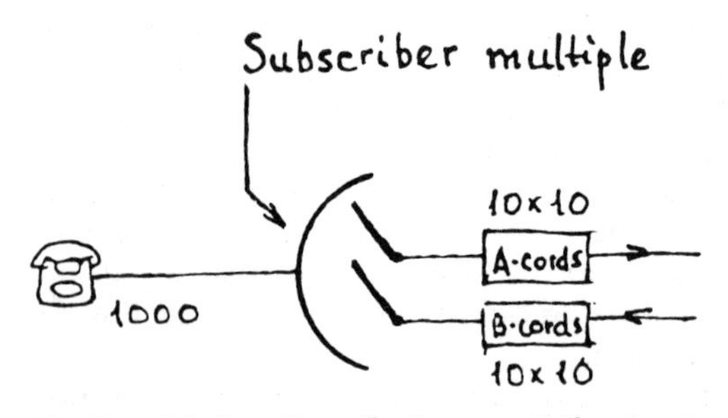

Fig. 10. Combining line-finders and final selectors.

In modern switching systems, the combined line-finders and final selectors are called a *subscriber switch.* And as we move away from the traditional selector into modern devices, we introduce a new name for the group selector. From now on, it will be called a *group switch.*

And with all this we have earned the right to simplify our diagram still further. See Fig. 11 on following page.

That really is all you need to know about switching at this stage. But you should know something about the brains that control the switches.

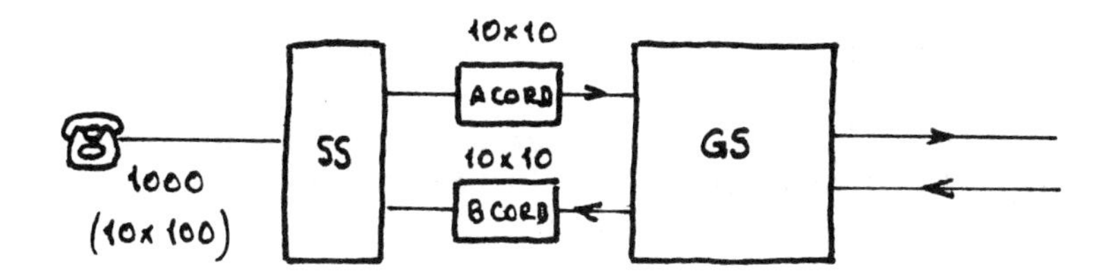

Fig. 11. A 1000-line exchange with 10 groups of 100 subscriber lines.

Everything under control

The last section concentrated on switching; we left control behind with the manual operator.

And it's true that in the advance to automatic telephony, the system did lose some refinement of control, though of course, it gained immeasurably in speed and economy.

Refinement of control is really a function of something like a brain, and the first machines to bear any marked resemblance to brains were computers. In the late nineteenth century, when telephony began to go automatic, computers were half a century away.

The first automatic switch was the Strowger, or step-by-step selector. For 60 years it dominated automatic telephony. In the Strowger system, the switches, or selectors, are activated directly by the pulses dialled from the telephone. The pulses from the first digit set the selector to a level which represents the digit, and then it rotates quickly to find a free outlet on that level. The next digit dialled sets the next selector to its level, where the selector rotates to find an outlet. And so on. Strowger is sound and reliable, and is still being installed, but it is cumbersome and heavy, and because it is activated directly by the telephone dial there's little room for a brain in the process.

The first L M Ericsson automatic system was the 500-point selector system. Unlike Strowger, it used *registers.* Registers act as a buffer store, receiving and storing the telephone number *before* the selector is activated. When it has received enough digits from the telephone, the register generates its own pulses to set the selectors.

A register system has several advantages. The Strowger switch is decadic—it can work on only ten levels, corresponding to the ten spaces on a telephone dial. With a register system, the selectors can be designed to operate on more levels. The registers can translate number combinations before they reach the selectors, which means that the selectors don't have to work through them solemnly one by one. A register system is more flexible.

In the 1950s, both Strowger and the 500-point system began to be superseded by a new system called crossbar—still electromechanical, but faster and, above all, more reliable.

Ericsson's crossbar system retained the registers, but added new pieces of equipment to set the switches, called *markers*. You'll note that for the first time in this section the word "switch" has replaced the word "selector". The distinction is a bit blurred, but within Ericsson we define a selector as a device which makes its own selection. It has a moving arm, and its own control built in, so that on its own it can test the outlets it passes over until it finds a free one and stops. A switch, on the other hand, in our language, is a passive device which depends for its setting on outside help, in this case, the markers. The marker does the testing, and determines the setting of the switch. For each call, it's used for only a fraction of a second.

Markers and registers together do form a control system of a sort. It's called common control, because it's quite independent of the switches, and common to all the operations of the exchange.

In a manual exchange, you will recall, the operator did two things. She did the switching—setting up and disconnecting the calls—with her hands. And she used her brain to receive information (register function) and to decide what to do with it (marker function). If a number was engaged, she would tell the caller. If it was free, she would set up the call. Nothing happened until she made it happen. She was a common control system—limited in capacity, but not in refinement or sophistication.

Common control systems received information and made decisions. To make things happen, they depended on a large number of devices called relays, which made or broke circuits on command. But by the end of the 1950s, the electronic digital computer had been developed sufficiently to look like an interesting alternative to relay systems. Relay systems had by now become immensely complex. International subscriber dialling and toll ticketing (the production of itemised bills) handled by registers and markers called many hundreds of relays into play for each call.

It was time to build brains back in!

By the mid '60s, a handful of computer-controlled switching systems had been designed. One or two of them even worked. But they were expensive, they were unreliable, and they were not easy to handle—modifying software, for example, was difficult and time-consuming.

Development continued, however, and by the end of the '60s, a few systems were actively marketed. Bell USA was the dominating force. At Ericsson, we had put in one computer-controlled exchange of a type we called AKE—of which more later.

Computers could advance only as fast as the electronic components they incorporated. These components, largely transistors and electronic memories, were available only as discrete components until the second half of the '60s. Many memory devices still used magnetic cores, which meant a great deal of fiddly, cumbersome wiring and associated expense. As a result, the designer of a system spent a lot of time and effort minimising the number of components to keep production costs down. All the tricks of the trade were employed to find multiple uses for memory, and to write programmes which required the minimum hardware. The outcome was some very complex systems—too complex, with hindsight, for widespread use, except, to some extent, in the American Bell companies.

The arrival of integrated circuits changed the whole picture. Component costs tumbled, and designers could optimise the total effectiveness of a system.

So what were these systems like?

In telephone switching systems, computers are called processors, and they control the switching process. Like any other computer system, they incorporate hardware and software, and the software consists of programmes and data. The whole technology of computer control is called stored program control, SPC.

These days, the telephony system they control also incorporates hardware and software.

The hardware you are becoming familiar with—switches; circuits—subscriber line circuits, trunk line circuits, cord circuits; and such things as signal receivers and transmitters. In an exchange with SPC, the hardware is very largely electronic; relays still turn up, but not in anything like the number there were in electromechanical exchanges. The switches may be of many different types—crossbar, mini-switches, electronic cross-point matrices, or digital. Don't worry about them. Just notice that word "digital."

The software is of two kinds: operating and applications software. The distinction is not difficult. If you buy a personal computer today, its functions are programmed into it in the operating system. But to avoid writing your own programmes to make it do something useful, you probably buy an applications package—a programme on a disc which will make it do word-processing, for example, or work out your tax. You type your data in and the applications programme tells the operating system what to do with it. It also operates the hardware to make something physical happen (a display appears on the screen, for instance).

Exactly the same thing happens in telephony. On a larger scale of course and with one very important difference—the processor controlling a telephone exchange works in *real time.* With a personal computer, or any large computer of the type we see in businesses and offices, the input is orderly; we feed the machine sequentially with one task at a time. In a telephone exchange,on the other hand, processing power is called for at random. We don't know at what moment a subscriber will pick up his phone to make a call, or when a call will arrive from another exchange. And several calls may arrive at the same time. The random nature of telephone traffic requires real time processing.

The control system is a data-processing system, made up of hardware, processors, and the *operating* software.

The telephony system is also made up of hardware—the switches and so on that we now know so well—plus the *applications* software.

The applications software is, however, stored in the memories of the control system, and executed on by the control system.

In Ericsson, we use the acronym APT for the telephony system, and APZ for the control system. Fig. 12 may help.

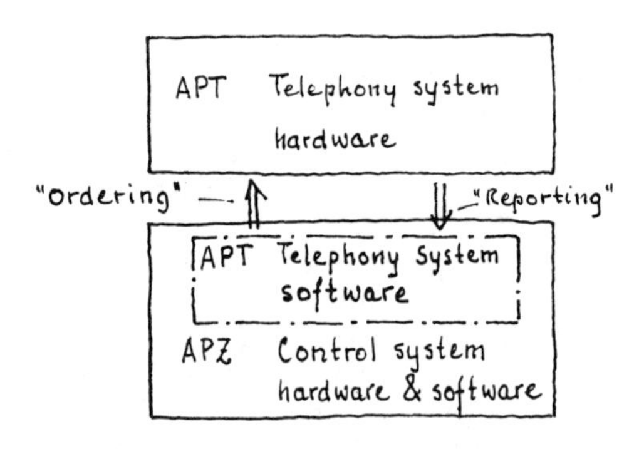

Fig. 12. Basic principle of a Stored Programme Control system

The principles apply to all SPC systems, but terminology and definitions vary within the industry. The diagram illustrates a principle. The actual structures — the architectures — of SPC systems vary greatly.

And now, at last, let's take a look at a telecommunications company which had played its part at every stage of this history of technical development. We'll begin at the point when the AXE story began, in 1970.

THE COMPANY OF LARS MAGNUS ERICSSON

Telefonaktiebolaget L M Ericsson (The L M Ericsson Telephone Company) is the parent company of the Ericsson Group. The Group is, or was in 1970, wholly dedicated to telecommunications. Its work includes the development, manufacture and sales of telephone, telex, and now data-switching systems, for public and private networks. It also covers private exchanges, telephone instruments, cable and wire, transmission equipment, radio and defence electronics, security systems, intercom, and components. Its headquarters are in Stockholm, Sweden.

But the Ericsson Group is truly international. More than half of its 75,000 employees work outside Sweden. Ericsson equipment is manufactured in some twenty countries in its own factories, or by other companies under licence.

The original company was founded in 1876 by Lars Magnus Ericsson, a young mechanic who set up a small workshop taking in telegraph instruments for repairs. 1876 was also the year in which Alexander Graham Bell invented the first telephone, and telephones were on sale in Stockholm within a couple of years. Bell telephones began showing up in Ericsson's shop for repair, and it soon occurred to him that he could make telephone sets himself which would be just as good. He brought out his own first telephone in 1878, and this was the start of L M Ericsson as a telecommunications company. The company grew quickly, and was internationally orientated from the beginning.

At the turn of the century, the largest Ericsson factory was in St Petersburg in Russia. There were also large plants in the US, and in England; and in Latin America there were a number of operating concessions. The first world war, and national politics, changed the pattern, and from the 1920s L M Ericsson gradually became primarily a manufacturer.

Looking back through company history, it is interesting

to note that in the 1920s L M Ericsson nearly missed the chance of getting into automatic telephony. Ericsson had been relying on manual systems, and was slow in developing automatic switching. Eventually, however, the first Ericsson automatic system came on the market in 1923. This was the 500-point system. At the time, its main characteristic was that it used registers, in which the dialled digits were stored, and which directed the setting of the selectors. It marked the beginning of the company's tradition in switching and the design of telephone exchanges.

However, it was not until the late 1940s, when L M Ericsson started using the crossbar switch, that the company became a major force in the international markets.

The original crossbar switch was invented and patented in 1913 by a man named Reynolds in the US. It didn't work. The design was improved and a system built in Sweden by the Swedish Telecommunications Administration, with exchanges put into service in two Swedish cities in 1926. The system proved too expensive and used too much space, so development was abandoned. But in the late '30s, rural telephony became an important issue in Sweden, and the crossbar switch was looked at again for use in small exchanges. In this context, it came into its own, and a range of small rural telephone exchanges was developed.

In the '40s, the US Bell System became interested in crossbar and developed its first system. And now, around 1946, L M Ericsson also got in on the act—the switch was perfected in collaboration with the Swedish Administration and a first system was developed and installed in Helsinki, Finland. Eventually, crossbar became accepted in many parts of the world, and L M Ericsson became one of the prime suppliers. Other suppliers, too, developed systems, and during the '60s crossbar became the main product within the industry. There were exceptions—the British stuck to Strowger, with the intention of making a single jump eventually into electronic technology; and in Germany, Strowger systems were developed even further—the EMD model was a wonder of electromechanical engineering, and was sold in large quantities.

Ericsson was living on crossbar, but other developments were going on.

In 1970, the L M Ericsson parent company was organised in several product divisions. A few departments and the manufacturing facilities formed a separate organisation. This meant that the product divisions were responsible for development, sales and administration, but bought their production from the factory organisation.

The parent company management had a double role, since it also acted as group management for the Ericsson Group. There were the usual management departments that one would expect, looking after personnel, accounts, control, legal matters, etc.

To avoid an extended written description, it may be easiest to look at a diagram.

Parent Company

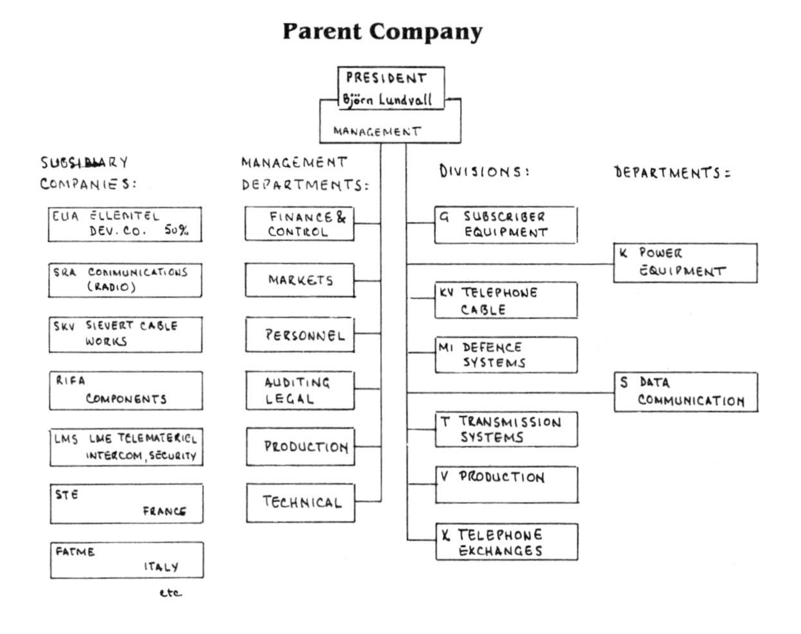

Fig. 13. The Ericsson Group, 1971

L M Ericsson is by nature and tradition an engineer's company, and in this context it's worth mentioning a second characteristic, which is not unique to Ericsson but typical of many large Swedish companies—people tend to stay. The combination means that you find many people in management with an engineering background. Like me, they may have started in the design departments and then moved over to marketing and sales and, in time, gone on up the ladder. Promotion is slow, by international standards—Ericsson differs from many US companies of a similar nature. In them, you will find much more movement among the top positions and, I suggest, less in-built continuity and tradition.

In this story, we are chiefly concerned with the Switching Division, or to use its official title, the Telephone Exchange Division. Within L M Ericsson it is referred to as the X Division. The letter X is often used to denote a switch or switching, as in British Telecom's current System X. In 1970, X had recently gone through a reorganisation and in diagrammatic form was structured as follows:

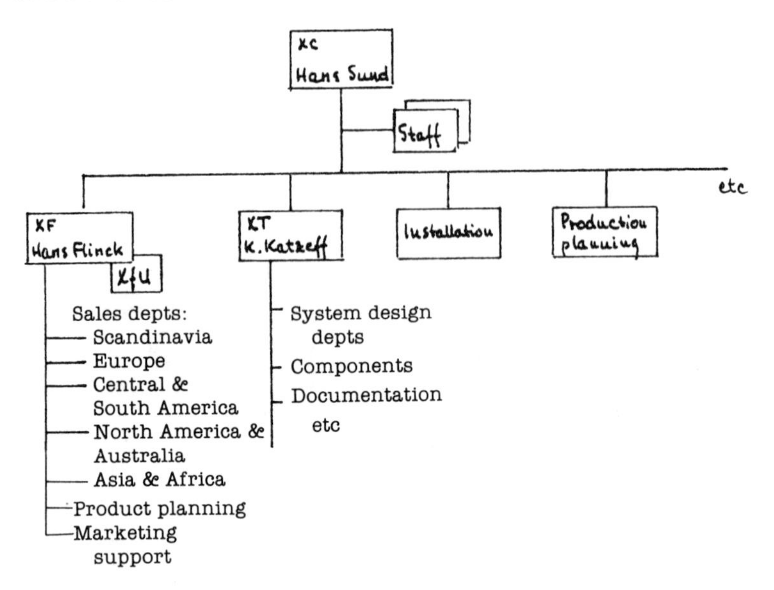

Fig. 14. Basic organisation of the Telephone Exchange Division

We had two major sectors, one for marketing and one for design and technical services; and in addition, we had several departments (such as installation and production planning) and some staff units assisting the head of the division (XC) in matters of planning, economic control and costing, personnel, and so on.

(Note from Jeans: XF is marketing. XT is technical. The C in XC stands for Chef. L M Ericsson is a stew of initial letters. The products are all identified by acronyms, and so are the people. And every letter is important. The latest PABX (private automatic branch exchange—not an exclusively Ericsson acronym as it happens) began life as the MDS 110. Later, it became the MD 110, *which was very important*. I still don't know why. When we began this book, John was DBC, head of corporate communications. All capital letters, you notice, *which used to be very important*, but isn't any longer. We do quite a lot of work for Lars Andersson, who looks after advertising for the X Division today. He is, or used to be, X/YxaC, *which was very important*. If you got it wrong when you sent an invoice, the paper used to wander round Stockholm with nobody to love it. Or pay it. You get used to it all.)

The head of the Telephone Exchange Divison (X) was Hans Sund. Hans had been in charge of the Defence Division (MI), and took over X from Fred Sundqvist. Fred had been made executive vice president of the parent company, in charge of all telecommunications operations.

Fred was one of us. He'd spent his whole career in X, and knew the telephone exchange business (and the Ericsson acronyms) from A to Z.

Hans Sund was different. He arrived in 1970 from MI with a reputation for a forceful style of management, and a hot temper. He had little background in telephony, but he knew his electronics. He had spent several years in the high-technology business, developing systems for the military. And although not everybody realised it at the time, the X division was facing a decade of dramatic and significant change. We were about to shift from electromechanics to electronics. We'd been working with electronics for many years, of course, but we were still relying on the old, tried-and-tested art of electromechanics for our daily bread and butter.

Head of the technical side of the division was Kurt Katzeff. Kurt had a long background in telephony and electronics, and had made major contributions to the advancement of technology. In particular, he was one of a pair of engineers who developed a fully electronic switch for the US Air Force, the 412L. (The design was transferred to North Electric in the early '60s.) In recognition, Kurt and his colleague, Gunnar Svala, received gold medals from the Swedish Institute of Engineering Sciences. Kurt had also been instrumental in planning and organising Ellemtel, which came into existence in June, 1970. But we shall not be seeing much of Kurt in this story, since in 1972, he suddenly decided to abandon ship and joined ITT.

On the marketing side, the XF sector was headed by Hans Flinck. Hans is a wise old bird (slightly older than me but a lot wiser, no doubt). He has spent his life in Ericsson, and in switching, and after some years in design and in Colombia, he has been active in sales and marketing. I became one of his staff of two, thus picking up a close working relationship that had begun back in 1954 when I went to Mexico and he was responsible for the Latin American markets. He had also been my boss during periods of the 1960s. Hans has always been a reliable source of counsel and guidance, and has been the ideal boss. He will not admit this, of course—indeed he makes a point of telling people that whatever he suggests Meurling ought to do, Meurling will do what he likes. This is, of course, all lies.

THE YEARS OF DECISION

The story of AXE begins, as so many stories still running in the '80s did begin, in the '60s.

The significant events of the '70s were yet to come—the oil crisis; the slow, painful recovery from the oil crisis; the world slump at the end of the decade. In the '70s, one concept—survival—increasingly dominated the thinking of every commercial organisation, every government, and practically every individual.

The '60s, on the other hand, had been a decade of growth. And, particularly, it had been a decade of growth in telecommunications.

The equipment for this growth, and in particular the telephone exchanges had largely been supplied by a dozen or so companies, a fairly stable rogues' gallery competing for the favours of the telephone Administrations—though by no means all the Administrations permitted such competition. The US was still closed to non-American suppliers of public telecommunications. Britain bought only from British companies. Other arrangements were less official but equally strong.

The value of investment in telecommunications had become so great that purely commercial or technological considerations were outweighed by political expediency, but even in the uncommitted countries, comfortable and continuing relationships had been established. In much of Latin America, for example, Ericsson would be aggrieved to find itself excluded from an Administration's tenders list.

But if the market carve-up seemed stable, there were nevertheless sharp differences between the suppliers. These were usually differences of preferred technology, and differences of perception of the ways in which telecommunications was likely to develop.

Who were these suppliers? And what were their products?

From Ericsson's point of view, the most significant was

the giant US corporation, ITT. A special relationship existed, and still exists, between ITT and Ericsson. In so many markets, Ericsson and ITT seemed to be the principal competitors. Part of the reason lies in history.

In 1932, the financier Ivar Kreuger, the Swedish match king, shot himself in Paris. Kreuger held the majority of the shares in L M Ericsson, and to Ericsson's horror, it learned that these shares had been deposited with ITT as security for a loan. The loan could not be redeemed, and ITT effectively became the majority owner of Ericsson.

Marcus Wallenberg, the Swedish banker involved in clearing up the mess after the Kreuger crash, played a heroic role in negotiating a deal with ITT. The deal left Ericsson intact to operate as an independent company, but ITT continued to hold a significant portion of Ericsson stock. It was 1962, 30 years later, before Wallenberg finally succeeded in buying out ITT.

There have been other situations involving Ericsson and ITT. Early in the century, both companies acquired operating concessions in Mexico, competing for the customers in the principal cities. In Mexico City, for example, there were two networks, and no connection between them. An Ericsson subscriber could not call an ITT, or Mexicana, subscriber. Full availability meant two telephones, one for each network. The networks were eventually interconnected, and finally, in 1948, consolidated in a jointly-owned company, Teléfonos de México. Ten years later, ITT and Ericsson sold out to Mexican interests—but both companies have continued as the main suppliers on the market, each with large manufacturing plants in Mexico.

At this point, it's necessary to pump in a little more technological description to make sense of what ITT (or anybody else) was about in the '60s. The details don't really matter—or if they do, we'll pick them up later—but the underlying principles are significant.

You'll remember that we divided the functions of an exchange into two broad areas—switching, the actual making of connections; and control. This distinction is much clearer in hindsight than it was at the time. Let's take it a little further.

We mentioned that in the Strowger selector there was no separate control system. And we said that Strowger was largely overtaken by the crossbar switch—still electromechanical, but more compact and reliable. In crossbar, separate devices called registers and markers operate the switch and provide the control.

During the '60s, other types of switching crept into the systems on offer. With one exception, they were still electromechanical, but they were further and further away from Strowger. Many of them will be mentioned in the next few pages.

We also looked at the coming of computer control— SPC. It was an interesting possibility at the end of the '50s. By the end of the '60s, practically everybody was offering it in one form or another.

The general picture of the '60s, then, is one in which most manufacturers were installing crossbar systems as fast as they could. There seemed no reason why demand for crossbar should ever dry up. But restless engineers were beginning to offer advanced alternatives, with apparently significant advantages and some competitive edge in at least some applications. And these switching advances were making stored program control more and more attractive.

For a while, and not for the first time, the suppliers were ahead of the market in their perception of possibilities.

At the end of the decade, ITT was typical. It was supplying its crossbar system (called Pentaconta) in great quantity. But it had also introduced the Metaconta switching system. Metaconta was a stored program control system which appeared in different versions. ITT's Bell Telephone Manufacturing Co. in Belgium sold a version called the 10C; the French ITT companies, CGCT and LMT, sold the 11A and 11R versions respectively. Standard Elektrik Lorenz, ITT's German arm, sold yet another version.

The differences were in the switches—3 pole reeds from Bell; miniswitches in the 11A; 2-pole reeds in the 11R; and so on.

The second main competitor was Siemens, of Germany. Alongside ITT's Standard Elektrik Lorenz, Siemens is the main supplier to the German Bundespost, and it has continued to develop the original Strowger, or step-by-step, switch. In Siemens' hands, the Strowger switch had become a very fine device—though still very much electromechanical. The last variant to come out was the EMD, in which the original noisy vulnerable sliding contacts had been eliminated. Around 1970, Siemens, too, was working on two systems with electronic control (more or less equivalent to SPC)—the ESK, with a crosspoint matrix switch, and the EWS, with a new crosspoint relay of the ferreed type. EWS was being developed in cooperation with the Bundespost.

A Siemens exchange is always neat, and the attention to mechanical detail and quality is impressive. (We call them the Mercedes-Benz of switching.)

The British companies were not very visible in the export field. They were busy making Strowger for the British Post Office. Plessey had introduced a crossbar system in the early '60s (5005), and later added a semi-electronic exchange (Pentex, or TXE2). STC, ITT's British company, was developing with the Post Office a wired-logic electronic control system. It later became the TXE4, which has been widely used in the UK, but not exported.

Philips had developed the PRX—an SPC reed-switch system, with a capacity up to about 10,000 lines.

From Canada, we were beginning to hear a lot about a new system from Northern Electric (later, Northern Telecom), the SP1. It had reed-switches and SPC, and Northern Electric was becoming interested in export.

Apart from ITT, the US companies were not, generally speaking, addicted to export. (Their home market in any case represents some 40 percent of the world's total.) Western Electric was designing the No. 1 ESS (ferreed with SPC), and was in many ways the world leader in technology. There were also GTE, Stromberg-Carlson, and North Electric—once an Ericsson company, and now developing a trunk exchange system, ETS-4, based on Ericsson's AKE (SPC, with code switch).

The Japanese telecom industry concerned with public switching was represented by four companies—Nippon Electric, Hitachi, Fujitsu, and OKI. These four had become very active in the world markets with the D-10 system, developed as a standard for Japan in co-operation with the Japanese Administration, NTT. D-10 combined SPC with a minicrossbar switch.

There were other manufacturers—in Eastern Europe, in the US, and, of course, in several other countries as subsidiaries of the large producers.

But only one other system need concern us. In France, the bulk of the lines being installed in 1970 was crossbar, made locally by Ericsson and ITT, among others. But one company, CIT-Alcatel, virtually unknown outside France, was beginning to attract attention at conferences and in its publications. Its system was called E10, and its SPC system was perfectly straightforward.

What was new with E10 was not the control system, but the switch. The E10 switch was digital.

THE ROOTS OF CHANGE

And within all this, what was happening at L M Ericsson?

The first Ericsson SPC system was AKE 12, and the first (and only) AKE 12 exchange was cut into service in 1968, in a small town called Tumba, in Sweden. The control system of AKE 12 was comparatively primitive and limited, but the Tumba installation was successful, and led to an early decision to develop a more powerful SPC switching system specifically designed for large national and international transit exchanges.

This improved system was called AKE 13, and the first orders were booked in 1968/69.

AKE 13 was to be second-generation SPC. AKE 12 used discrete semiconductor components. AKE 13 used integrated circuits, and because it was intended for large applications, the control system was of the so-called multi-processor configuration, with up to 8 duplicated central processors. The maximum capacity of the system was about 64,000 trunk lines. AKE 13 used code switches—a sort of souped-up crossbar.

At that time, the late '60s, I was heading a department responsible for telephone exchange sales to North America and Oceania. My most important markets were Mexico and Australia—where crossbar exchanges were being made at the Ericsson plant in Melbourne (Ericsson Pty Australia), and under licence by Plessey and STC in Sydney.

In 1968, the Australian Post Office (as it was then) issued an international invitation to tender for a large transit exchange, Pitt, in Sydney. This was precisely the sort of application for which AKE 13 was designed, and we spent several months with Ericsson Pty putting a proposal together. We based the proposal on AKE 13 because the Post Office had identified many requirements resulting from the fast growth of national subscriber dialling, and such requirements were obviously best met by applying computer control. The only other serious contender was ITT with Metaconta. We handed in our offer in January 1969 with high hopes.

In September, the Australian Post Office announced that they had chosen Metaconta. This was a serious blow, which I took rather personally, and which was to have a major effect on the later course of events.

But the SPC efforts continued at Ericsson. Development had begun as an investigation into the feasibility of replacing the logic in crossbar registers—at that time nearly 100% based on relay circuits—by electronics. The results had been disappointing, and had indicated that we should only get significant upgrading of features and performance by using stored program control. Most notably, SPC would give us a new level of operation and maintenance. Work began on the introduction of SPC into the ARF and ARM crossbar systems to transform them into ARE 11 (for local exchanges) and ARE 13 (for transit applications).

A third element in the development impetus was contributed by Mexico. The Mexican Telephone Company (Teléfonos de México, my former employers) had identified specific needs as a result of the rapid expansion of the Mexico City network, and the growth in automatic long-distance traffic. In Mexico City, the network was growing more dense—the growth was in number of subscribers per square mile rather than in the growth of the area covered by the network. This development results in very large exchanges. It then becomes most economical to handle traffic by setting up a lot of routes direct from each exchange to all the other exchanges, rather than taking it through intermediate tandem exchanges.

In this environment, the limitations of our existing crossbar system began to show up. The group switch was not large enough, and its access range was limited. In discussion with Telmex, we first designed an additional stage in the crossbar group switch of ARF, and later developed a completely new group switch, called ANC 11. ANC 11 could be incorporated into the crossbar environment as a local exchange, or could work as a separate tandem exchange. ANC 11 used electronic control, and, at this stage, the code switch.

The situations in Australia and Mexico could be paralleled in many of our markets. No definite large market for SPC had appeared in our spheres of interest, but external market developments and internal restlessness in Ericsson were building up the pressures for change.

It was time for a review.

What had we learned from the many years we had put into stored program control technology?

Well, we had learned to design processors and software for SPC. We had developed a construction practice that we could implement and test. We had built a great deal of testing equipment for production and installation, and we had trained several hundred engineers—our own, and those of our customers.

But most important, we were beginning to feel that SPC, as we had it, and indeed as our competitors had it, was a complex and expensive technology. We began to think about *handling*—ease of installation, support, manageability. We were convinced the technology was possible, and indeed necessary. In the few specific applications with which we had been involved, the need for the functions and features which only SPC could provide had been very pressing. But those applications had been at the level of major national and international transit centres. Could the extra cost of SPC ever be attractive in local exchanges dispersed in large numbers over wide areas?

We could sum up like this.

The technology of electronic circuits was developing rapidly—memory cost, in particular, was dropping dramatically. The costs of handling—*all* forms of handling, design, testing, modifications, fault correction, production, installation, and operation and maintenance—of our existing SPC products were, we felt, unacceptably high for general application. And we had no real competitive edge in our products. In certain competitive situations, we were finding that even our switches, crossbar and code, were regarded as slow and passé.

It's not surprising that among some of us ideas were beginning to germinate. Ideas not just for a new control system, but for a new switching system, based on new concepts.

And we were not alone. The Swedish Telecommunications Administration, Televerket, was having similar experiences. Parallel to Ericsson's AKE 11 programme, Televerket had designed a system called A 210, which went into service as a large local exchange in the Stockholm area in 1968. But Televerket, too, was reviewing its experience, and ideas were forming for a new system.

One other aspect of the situation was of concern to the top managements of both Televerket and L M Ericsson. Sweden is a small country and represents a limited market. It seemed obvious that to continue to develop competing major systems was wasteful. The number of engineers available for developing a switching system was limited, and the design of a new system would require considerably more manpower than anything previously undertaken by either organisation. It was natural to look for some way for the two organisations to cooperate.

Björn Lundvall, president of L M Ericsson, and Bertil Bjurel, director general of the Swedish Telecommunications Administration, set in motion a series of studies and discussions. In June 1970, a jointly-owned company was formed. Ellemtel (full name, Ellemtel Utvecklings AB) would be a company that would undertake design and development projects exclusively for the two parents, separately or combined.

Among the first tasks defined and placed with Ellemtel was a project called AX, the development of a proposal for an SPC local exchange system which would offer positive cost benefits.

CHAPTER 2: BIRTH OF A WORLD-BEATER

THE REQUIREMENT SPECIFICATION

IT'S important at this stage to get a few distinctions clear.

On AX, Ellemtel was not being asked to start product development. Ellemtel was being asked to produce a *proposal* — a paper description of the way in which Ellemtel's engineers would tackle the solving of the AX problem, and what the solution would consist of.

It is, of course, nonsense to set about producing solutions to problems which are not defined. Before you start designing a telephone system, you must (or at least you should) spend some time putting down on paper all the things you want this system to be able to do, what it is to be used for, what the acceptable cost will be, and so on. Such a document is called a requirement specification, and in our case it was to be one of Ericsson's instruments for steering the system-development program at Ellemtel.

The development of our first SPC system, AKE, had been managed almost exclusively by Ericsson's technical departments — in fact, it started life in the hands of a small group of technical wizards. They were excellent people, but with limited contact with the markets. These people had written their own requirement specification, with emphasis, we felt, on technical design criteria, rather than on the practical needs experienced by marketing and sales people in daily contact with customers and their problems.

So it was decided that the L M Ericsson requirement specification for AX should be "written by the markets." In practice, the task was given to the marketing sector of the switching division. To do the job, a small working group of five was formed in the autumn of 1970. I was one of the five.

Of course, sales and marketing staff in the X Division of L M Ericsson were not foot-in-the-door, fast-talking, commercial travellers. They were, and are, engineers — not design specialists, but people with a deep knowledge of technology and great experience, acquired from close association with telephone Administrations around the world. I spend my first years at Ericsson as a design engineer on crossbar systems, and then spent eight years on the technical staff of Teléfonos de Mexico. My colleagues in the group represented several other disciplines in telephony. They possessed a thorough knowledge of development trends within subscriber services, operation and maintenance, traffic requirements, reliability engineering, and so on.

Our job was to collect facts, ideas, concepts and experience into a cohesive and thoughtfully edited volume, which would provide an up-to-date description of our ideal system. A practical ideal, that is, not a dream product as an unattainable goal. Our engineering background would lead us to formulate requirements that could be met within cost limitations. Our marketing experience would lead us to make these cost limitations realistic. And our enthusiasm would make us do all this in four months!

Well, a preview edition was available in January 1971, though the completed and approved requirement specification actually appeared several months later. Ellemtel responded with a draft proposal, to which we responded with a series of modifications, and so 1971 slipped by.

Arguments were often heated, criticism was frequently scornful, fraternal greetings were often no more sincere than those between two Iron Curtain countries. But the AX concept progressed very steadily.

From the welter of detail and the solutions to a series of fragmented problems, some broad principles began to emerge.

Early on, of course, Ellemtel confronted the problem of having two, different, requirement specifications! Televerket's version naturally reflected its vision of the Swedish telephone system, and how it would grow and develop. L M Ericsson had world-wide export in view.

Two reconciling principles eventually appeared, both of them to exert great influence on the new system as it developed.

First principle—a modular approach

The first of these principles rested on the global nature of the Ericsson specification, which tried to cover *all* telephone exchange markets. Naturally, that would include the Swedish market, which should be able to be fitted within the larger framework. The trick was to do this without a cost penalty—without making the end product unnecessarily expensive for Sweden by building in features which would not be used.

The obvious solution was to build up the system in such a way that features, or feature packages, were designed in blocks—modules—that could be included or left out. This modular principle was essential to Ericsson, which was looking for a system which could be sold to a wide variety of markets, with a wide variety of requirements.

A case in point is traffic capacity. After much argument, the two parties agreed that the system should be designed with four versions of subscriber switch with characteristic traffic-carrying capacities of 0.24, 0.16, 0.12 and 0.08 erlangs respectively. (At the time, we were basing our concepts on different types of electromechanical switches, the so-called mini switches or reed relay matrices, and each subscriber stage version represented a different switch combination.)

Televerket put a lot of effort into the requirements for operation and maintenance, drawing on their vast experience as an operating organisation. They paid particular regard to the concept of defining the administrative functions of telephone exchanges—functions such as charging, handling subscribers, maintenance, fault-reporting—as sub-functions within the total administrative system of the telephone operating company. These early concepts later evolved into a series of feature packages of great sophistication which became one of our most important marketing points.

Second principle—ease of handling

We were tackling the problems of handling, and we were beginning to play with new concepts. These concepts sound obvious today, but at the time they caused a great deal of discussion. Eventually, they crystallised in the specification as a series of requirements for "ease of handling." Some could be quantified in terms of man hours or running time. Some could not.

The system, we said, must be *easy to install*—it should be possible to reduce the installation time in the field for a 10,000-line crossbar exchange from the typical 12 months to half that figure. The system must be *easy to maintain*—it must incorporate functions which could detect and pin-point any possible faults rapidly and reliably, and it must have mechanisms built in which would stop a fault from causing secondary faults and degradation of the service.

And so on—the system must be easy to design, in both hardware and software; easy to manufacture and test; easy to document; easy to teach (to maintenance staff, for example).

In my own mind, I was beginning to realize that any system which could live up to all this would have another valuable characteristic. It would be easy to sell.

Why be small-minded?

We had originally been concerned about the cost of SPC in local exchanges. This was why our original brief had specified AX as an SPC local exchange system.

At Ericsson, we were beginning to have second thoughts.

How, we asked ourselves, would an Administration look at the AX concept that was emerging? There was no doubt that the ease of handling and manageability would be attractive— but would any Administration be interested in a system which, however excellent, could be implemented only in local exchanges? If not, these revolutionary advantages would be wasted.

The question led us to define a product with much wider applications—not only in local exchanges, but also in tandem and transit exchanges.

At the same time, we also saw a major opportunity in the concept of the remote subscriber switch. If a local exchange could be installed in the network with some of its subscriber stages "remote"—placed, for example, in transportable housing—it could mean significant savings in the cable network.

The idea led to further thoughts on the structure of the system. A local exchange has two basic switching elements: the subscriber switch; and the group switch, or a group selector.

You will remember the exchange diagram (shown here again).

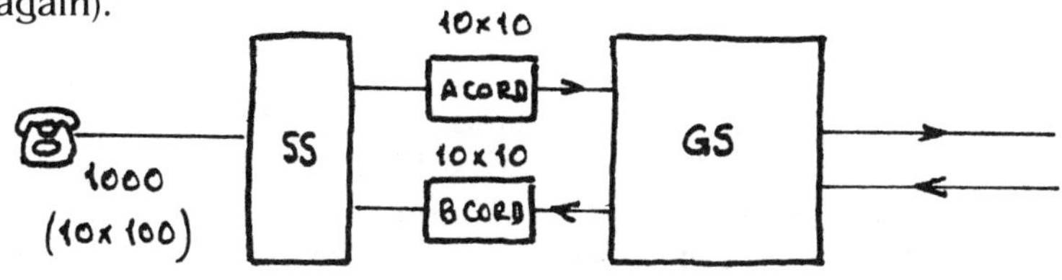

Fig. 15. A 1000-line exchange with 10 groups of 100 lines.

A tandem exchange is only a group switch, plus the necessary control and trunk circuits. A transit exchange is also a group switch, though it usually requires the switching of four wires instead of two.

There was no reason why the group switch should not be separated from the subscriber switch, and become a discrete unit. Indeed, we were beginning to conceive of the local exchange system in this way, made of two separate and more or less autonomous parts which we called the subscriber switch subsystem and the group switch subsystem.

The concept threw the spotlight away from SPC, and straight back onto switch technology. Initially, we were planning to use electromechanical switches (reed relays or miniswitches), but as the cost of electronic memory circuits dropped, we began to see that digital switches would one day—in the near future—be feasible. And a digital group switch in particular would offer great advantages to Administrations by cutting the investment required in a network.

It would be absurd to start all over again in a couple of years' time and develop a new, digital, system. AX would have to be designed to absorb digital switch technology as soon as it made economic sense. The subsystem concept would make this much easier—replacing a switch in an autonomous subsystem would not affect the rest of the system, and any upgrading could take place smoothly. (In the end, as you will see, the digital group switch was introduced into service less than 18 months after the first AXE cut-over, and the digital subscriber switch followed in the early '80s.)

So where had we got to?

By the middle of 1971 we had produced a requirement specification, shared by L M Ericsson and Televerket, which described the features and characteristics of a new system.

It should be an SPC system, providing a host of add-on subscriber services; and it should offer an extensive menu of operation and maintenance functions.

It should be suitable for exchanges with up to 40,000 subscribers and a typical traffic capacity of 0.10 erlangs with a 100-seconds average call-holding time. (This translates into a call-handling capacity of 144,000 BHCA.)

It should be able to be implemented in all types of telephone network applications—local, tandem, and transit exchanges—and should allow remote subscriber switches.

The system should provide ease of handling in all phases of development, installation and operation. Its structure should be such that new technology—notably digital—could be introduced, and new functions could be added as market requirements emerged. What we were describing here is sometimes referred to as an open-ended system, or a future-proof system.

And we were asking Ellemtel for a proposal of how such a system would be designed.

THE DECISION

So by the beginning of 1972 we had a possible system very clearly defined, and Ellemtel had produced a proposal with a plan for a development program.

But describing a requirement is one thing. The decision to implement it is something quite different. Our requirement specification was a major advance, in that we were quite sure that it represented at least latent market desires. Nevertheless, a decision had to be made, whether or not to go ahead with the AX proposal.

The question was this. Should we spend 50 million dollars, or 1200 man-years, on developing a completely new switching system (AX)? Or should we up-grade and develop our existing highly-successful crossbar system (AR)? Or should we put more effort into our existing SPC system (AKE)?

The problem was not merely one of cost. Time was also of the essence—with the resources we had available it would take five years of development before the AX system would give us even one exchange in service.

Having worked out the proposal, Ellemtel naturally wanted to go ahead. So did Televerket.

But at Ericsson, we were not single-minded. Any switching engineer of mature years has, if he's worth his salt, a new switching system up his sleeve. Some of these were on offer as alternative proposals alongside AX for evaluation.

In their favour was the fact that they would take much less time and effort to develop than AX. Against them was the fact that they were based on conventional structures, and did not offer any significant new features.

The head of X Division was Hans Sund. The final decision would be his.

We took a final look at the market and the competition.

ITT was marketing Metaconta with some success. The Japanese were marketing D10. Philips had its first contracts for PRX; and in the US, Western Electric was supplying the Bell System with the No. 1 ESS system.

We were selling AKE for transit exchanges, in particular for international exchanges. We had lost to ITT an order from the Australian Post Office, but we got some consolation out of getting from Australia's Overseas Telecommunications Commission (then a separate body alongside the APO) the contract for an international exchange in Sydney (1971).

All these products incorporated analogue switching. The bulk of the equipment sold by us and our competitors used straightforward crossbar, reed relay, or some similar electromechanical switch technology.

The true SPC market was actually small, although the new systems were heavily publicised, and much talked about at conferences and symposia.

But in France, CIT-Alcatel had introduced the E10 system, and E10 was digital. E10 was initially to be installed in rural areas and was designed for small applications. In our league, E10 had limited capacity and appeal, but some extremely interesting facts were coming out of France. Most notably, the combination of digital transmission with digital switching was opening up significant possibilities for reduction of the investment needed in telephone networks.

Plenty of people were studying digital switching, of course. We were doing theoretical work in Stockholm, while our Australian company had taken the design of a digital group switch to the working laboratory model stage.

Of the three associated technologies—digital switching, digital transmission, and SPC—SPC seemed pivotal in the decision. Could we see any telephone Administrations making decisions to introduce SPC technology—and if so, when? And should we be competitive over the next few years with the products we already had, or could improve by some further development?

The following table sums up the situation. It lists the main Ericsson alternatives and what was known of the competition. The listing is in terms of local exchanges, the application which would best represent the volume of the future market.

The table on the next page dramatises our difficulties.

We had ARE, but its competitive strength against systems using fast, miniature switches (reed-relay matrices or mini-switches) was doubtful.

We had an early version of AKE. This could certainly be improved by development based on the transit exchange design in progress at the time, and could result in a new product by 1974. But we knew already that AKE carried a penalty in the cost of handling. AKE, too, would probably be uncompetitive.

With AX, on the other hand, we should have a superior product, a system which—and of this we were certain—would be able to compete very successfully. But we should not have a new product in service until 1976, and the development costs would be at least four times those of AKE.

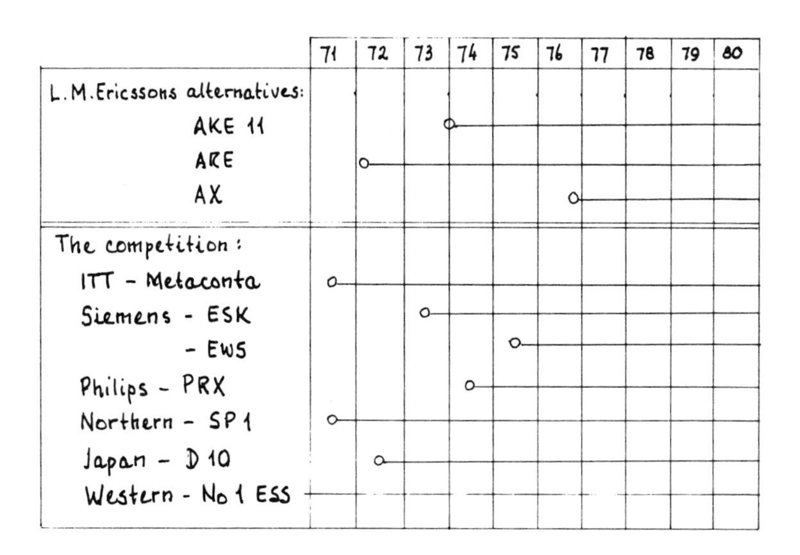

Fig. 16. Ericsson's system development alternatives relative to competitors' products, as seen in early 1972. A ring denotes first exchange in service.

This book is called A Switch in Time. The title reflects the character of the decision we had to take, to my mind a decision which would make any technical decision trifling. All our major competitors had their SPC products either in the market or near completion. How many markets would they have seized before we were ready? Would there be a significant development of markets for SPC local exchanges between 1972 and 1976, a period during which we would have nothing to deliver but promises?

Any decision to go for AX would be a gamble, though perhaps not quite as reckless a gamble as this brief description of the situation makes it sound. Hans Sund and a small group around him were convinced.

There was, however, a strong lobby arguing that since crossbar would be the most popular product for many years to come, there was no urgency about bringing out a new system. It would be better to improve our crossbar production resources. Even Televerket was not in any great hurry at the time. Hans had to convince not one, but *two* groups, each with different standpoints.

The decision was taken around February 1972. The AX switching system *should* be developed along the line proposed by Ellemtel, with a target date of 1976 for the first pilot exchange to be in service.

Some work had been done on a local exchange version of AKE 11, and we even had a couple of orders for this system. Work on it was stopped; Ellemtel resources in public telephone switching systems were to be concentrated on AX, under the leadership of Bengt Gunnar Magnusson.

GETTING THE MESSAGE

So far, we've looked at the story of AXE largely from the marketing point of view.

Marketing men tend to do this. Most of the obvious drama in the life of a commercial company is necessarily commercial. But engineers make the proposals. And decisions are no better than the proposals behind them. It's time to look at AX from an engineering point of view.

In August 1970, Ellemtel had been given four tasks—to develop a new subscriber exchange, a PABX; to produce a study on the building of a Swedish data network; to work out a proposal for a new public switching system, AX; and to study, evaluate and make proposals on new switch principles and designs.

Ellemtel was organised into four main departments, corresponding to the four projects. We, of course, are interested in the work of the public switching department, Department X, but it may be worth glancing at the results of the other departments.

The PABX appeared, and had a successful, though short, life.

The data network now covers the whole of Sweden, Denmark, Norway and Finland. It's ahead of everything else in the world—perhaps too far ahead. It matched the requirements of the four countries so perfectly that it's not easy to get the system installed anywhere else. This was a trap we were determined to avoid for AX.

The first new switches—reed switches—incorporated in AX were Ellemtel designs, and Ellemtel later took on board the digital switches for AXE which originated in Ericsson's Australian company.

The head of X Department was Bengt Gunnar Magnusson, with a steering committee to guide him. This product committee—PK-TS—still exists. It included three members each from Ericsson, Televerket, and Ellemtel. Its chairman in 1970 was Erik J. Eriksen, managing director of Ellemtel.

The product committee had an important role. It followed progress step by step, studied proposals as they came up, suggested the content and direction of work to be undertaken, and reported to a coordinating committee which approved suggested projects and made funds available.

It may sound dauntingly bureaucratic, but PK-TS played a very significant role in steering AXE to become the product it is today. We shall hear more of it.

In 1970, Bengt Gunnar and his bunch began by evaluating the proposals for a new switching system that were already knocking about.

There was A210 from Televerket, an SPC crossbar system developed during the '60s, and a variant called A210 Ux, intended as a maintenance centre.

Within Ericsson, we had AK69 and AK70. AK69 was based on functional modules, each with its individual control system, while AK70 was based on a central processor. There was even LX70, an attempt to combine the characteristics of both AKs into a single system which could be supplied in different versions for different applications.

While Ellemtel was engaged in this preliminary evaluation, the two parent companies were at work on their respective requirement specifications.

PK-TS, meanwhile, got involved in a heavy discussion on the advantages of logic in hardware compared with logic in software.

This was the notorious SLIC and SLIM controversy—System Logic In Circuits (in other words, reproducing the traditional electromechanical circuitry in electronics) versus System Logic In Memory (thorough-going computer control).

Since SLIC needs less memory capacity, it's true that it made economic sense while processors and memory were fairly expensive, but the attraction of SLIM was the flexibility of stored program control; the hardware can be engineered to do a variety of jobs if it's supplied with different software. Eventually, SLIC faded away, and we all became fully committed to SLIM.

By the end of 1970, Ellemtel had completed a study which led to the proposal for a system it called AX—in two versions: AX-N for exchanges with up to about 20,000 lines; and AX-H for high-capacity applications.

AX-N was based on a concept of functional processors—a form of decentralised control. It was a bit complicated. And to purists, it represented a compromise, not exploiting the full potential of computer control.

AX-H, on the other hand, had a structure of a central processor, plus signal processors.

At Ericsson, we weren't all happy with the Ellemtel proposal. There were various reasons for our concern.

One, I'm afraid, was the famous NIH (not invented here) syndrome. Some of our engineers felt that given our long experience of international markets we were better equipped to design a new system at Ericsson. Some of them had their own design proposals.

But there were some more serious reasons for doubt, and we decided to put together a study and evaluation group, which was given a month to document our objections. Björn Svedberg, President of L M Ericsson today, chaired the group, and it set to work to produce a report. We split into subgroups, each group tackling different aspects of the AX proposal.

We used our own earlier systems as reference points, and we spent a lot of time evaluating our competitors' systems.

In many ways, the report was a reappraisal of our requirement specification. It extended and elaborated on the governing factors we had arrived at earlier.

It focussed again on:

— handling costs, which for our earlier systems were far too high;
— network development trends, which suggested that a new system should offer a structure with separate and fairly autonomous group and subscriber stages;
— the variety of markets, and the need to adapt to them;
— the marketing importance of sophisticated operation and maintenance functions;
— the decision that all complex functions should be implemented in software (SLIC was dead!);
— the crucial importance, therefore, of programming aids.

These considerations seem obvious now. At the time, they were new, if not exactly revolutionary. As we hammered away at them, they had a profound effect on system design.

I should say here and now that the AX proposal did indeed incorporate most of these characteristics. What our evaluation did was confirm that we were on the right lines.

Although not everybody was happy, Ellemtel was directed to continue along the lines proposed.

We had, however, identified some points which needed further work. In April 1972, we presented a new report. We were worried about:

— cost, which would be too high for small applications of the system;
— capacity, which the Ellemtel proposal had indicated could not exceed 32,500 subscribers (compared with the 40,000

we had specified, at 0.10 erlangs per subscriber);
— reliability, which we thought was doubtful.

Since we had now refined and concentrated our nagging reservations under these three headings, and since cost, capacity and reliability were the marketing cornerstones of AX's later success, it's worth considering them in some detail.

Our concern over cost was largely rooted in history. We had analysed the sizes of the crossbar exchanges we had supplied over the last 20 years or so, and we had found that the average size for an initial exchange installation was about 2000 subscriber lines. At that level, the AXE proposal seemed too expensive compared with crossbar.

But two factors eventually persuaded us to accept the proposal.

The first was the very clear trend in component prices. Of course, our pricing had taken this downward trend into consideration, but further careful study showed that we were still being too conservative.

The second factor emerged from the network studies we were doing.

On the one hand, we could see that many orders would be for replacement exchanges. These exchanges would take over from existing installations which were either worn out or obsolete, because the services they could supply were limited. Such exchanges, often originally installed in the '20s and '30s and later extended piecemeal in installments, were usually city exchanges—the city exchanges were where automatic telephony began. They had turned into large exchanges, and if they were to be replaced by AX, the orders would be for large exchanges. When we studied the order pattern in our typical large markets, instead of over the whole range of initial installations, we found that the average size to be expected was likely to be around 10,000 subscriber lines. At that level, any calculation of cost per line produced a much more realistic cost for AX.

On the other hand, we would have something new to offer to parts of the world where there are no, or few, telephones—rural areas, particularly. Third-world countries are important markets for Ericsson. Here we found that the solution could be provided by digital network technology. We've described the principles of the remote switch, in which a portion of the exchange (the subscriber switch) is installed away from the main exchange, nearer to the centre of gravity of the subscribers' locations. From a switching system point of view, this meant that a network for a rural area would no longer consist of one main exchange and a number of small self-contained exchanges; it would consist of *one* system—the main exchange in a country town, and remote switches in the villages around it. Such a system would be comparatively large, with wholly acceptable costs per line.

So much for cost. Our reasoning (which turned out to be correct) strengthened even more our insistence on a system with sufficient capacity for large exchanges. We had specified 40,000 subscriber lines with a guaranteed traffic per line of 0.12 erlangs—144,000 busy hour call attempts. Traffic-capacity calculations for systems which exist only on paper are not easy. They involve adding up long streams of component operating times, measured in nano-seconds, within a model representing a traffic mixture of different types of calls. Safety margins must be included—and they also add up. We were on uncertain ground—and would continue to be until we had some real processors operating, and could load them with real or simulated traffic.

Ellemtel responded by speeding up the processor, and we accepted the design proposed. Three years later, we measured handling capacity on a working processor, and confirmed that it really did have the required capacity for 40,000 subscriber lines, with a good margin.

Any piece of telephone equipment has to offer certain functional guarantees, which are expressed in different forms. One set tells the customer how often he should expect the exchange to develop a fault, so that he can work out how many maintenance staff he will need and how many spare parts. This sort of reliability is expressed in such figures as "mean time between failures", or fault frequency. They are fairly easy to calculate with a well written program and a large computer to run it on.

Another set of guarantees covers what the customer's customers will feel about the service they're getting. It is called availability, and can be expressed as the probability of his call getting through at any time. Availability is a composite of a number of loss factors—loss due to lack of circuits, loss due to faulty equipment in outside plant and exchanges, loss due to inadequate transmission level or noisy transmission, and so on. We think 99.98% availability is about right. Anything better costs too much. Anything less means loss of revenue, and subscriber complaints.

With all types of telephone exchanges under common control, the question of what happens if some or all of the common equipment fails is important. With the central processor proposed for AX, *all* connected subscribers would be out of service should this processor fail. For this reason the central processor is duplicated, and the probability that both sides would be out at the same time is very low. An industry design standard to which the major manufacturers adhere is a maximum down-time of two hours—during 40 years. Here, AX was entirely acceptable.

Our concern about reliability focussed on the regional processor. We were afraid of the outcomes of any malfunction in it. (Regional processors are the smaller, comparatively simple-minded processors in the system, which perform repetitive routine functions for a group of lines.)

But the AX system proposed was engineered in such a way that if a regional processor failed it would affect a full group of 128 subscribers, or 32 trunk circuits. This, we felt, was cutting things too fine. As always, we tried to work out how good the outage figures of competitive systems were. There were, and are, no industry standards, but the major companies had arrived at certain recommended levels, and we felt we were falling short a little. After prolonged discussion and studies of various different solutions, it was agreed that regional processors, too, should be duplicated.

All these points were accepted, and a full development programme was placed with Ellemtel.

We were off! And high time, too—Bengt Gunnar and many of his Ellemtel colleagues had been increasingly irritated by our slowness within Ericsson. They were itching to get on with the job, and couldn't understand our concern over a lot of detail.

But we had been, and still were, genuinely concerned. In 1973, we made a third study. Again, we questioned cost, capacity and reliability, but with the project well under way, we had developed a way of working together without friction, and this study aroused less acrimony.

At this time, early 1972, I was appointed "object leader" for AX. This was an inventive title for a rather special responsiblity. I understand that Hans Sund had introduced this project working method during his time with defence systems, and saw similar needs with AX.

As object leader, my responsibilities were, briefly

- to monitor the project, flag up any deviation from the planned time schedule, discuss the consequences of any deviation, and make suggestions, with the project leader and the marketing team, for any corrective measures;
- to monitor product cost (which became increasingly important when we started making proposals to customers);

— to monitor the cost of the design programme;
— to coordinate the initial marketing activities;
— to make sure that the requirement specification was being fulfilled;
— to initiate the production of sales documentation.

The functions, at least, were important. In the excitement of product development, points like these can easily be overlooked. It's an object leader's job to keep an eye on the basics.

Developing system criteria—and a relationship

Broadly speaking, AXE has been through four crucial stages. The first was a sort of creative dissatisfaction with our existing systems. It was made more acute by such events as the setback in Australia, and the criticisms of Telmex. It culminated in the decision to establish Ellemtel, the preparation of the requirement specification, and the decision to develop AX.

We were now in the second stage—creative system design. If it failed, the third and fourth stages (creative marketing and consolidation) would not be required!

Ericsson, Ellemtel and Televerket have excellent engineers—but so have many other companies. The art is to get all these engineers working in an original and fruitful way, so that a world product emerges—at the right time.

You will have gathered already that there were plenty of people eager to say that we were getting it wrong.

There was the NIH syndrome. There was the necessary secrecy, which encouraged ill-informed criticism. There was doubt about our ability to maintain the secrecy anyway. There was the problem of how much Ericsson or Televerket should interfere with Ellemtel. And so on. Engineers, after all, are human.

But we had a big, difficult, urgent development project to run. We had little time for philosophical theorising about our working methods. And with hindsight, I believe we got two things dead right.

Most of us agree now that the decision to form Ellemtel and give it AXE to design was brilliant. The feeling of challenge that develops automatically in a new group in a new company with a new project is terrific—far greater than anyone can generate by adding one more code-named proposal to a long existing list on which an existing group is already working.

A feeling of challenge is essential to create group responsibility, and acceptance of a common, urgent and important goal.

The methods may well be unorthodox, but a small group can accommodate the unorthodox. Its management can allow unconventional individuals to work in unconventional ways. Goran Hemdahl was one such unconventional individual. If Goran was not a genius, he was very close to it. Many of his great thoughts would come to him as he paced the corridors of the Ellemtel office. Everybody knew this, and Goran was left alone to pace. Goran did a lot of the important groundwork for AXE—he devised a revolutionary software architecture; introduced the concept of modular software; and worked with the high-level programming language. When we started to market the system in 1974/75, I worked with Goran in writing about the AXE software. He must have found me stupid and slow to grasp his finer points on the handling of software. What saved me from being thrown out was probably the fact that we are both interested in certain types of jazz. In 1977, there was a ghastly misunderstanding, and Goran felt that his contribution was not getting proper recognition. He was snapped up by ITT to work on System 12, in Brussels, but has since left ITT as well.

The second thing we got right, I believe, was PK-TS, the steering committee. It provided the right blend of autonomy and mutual intimacy with the project, and it allowed a genuine three-way exchange of views between people who knew what they were talking about. There was no feeling that Ellemtel was reporting to superiors, or that Ericsson or Televerket was issuing orders.

The fact that the project was assigned outside the two parent companies had at least one other very salutary effect; it forced both sponsors to formulate the project properly and to be very specific. Ellemtel asked us both to put everything down on paper, hoping that they could then close the doors and get on with the job in peace and quiet. This method wasn't, of course, really workable. We limited the number of people involved to the minimum for the first few years, but obviously L M Ericsson had to have day-to-day contact with the project. It's more or less impossible to write a comprehensive requirement specification once and for all, and new problems kept cropping up which had to be solved jointly. And once we started to talk to customers, we needed constant information and support from the design engineers.

I joined PK-TS late in 1972. My role was primarily to look after the interests of the X Division marketing organisation.

Bengt Gunnar was suspicious of me to begin with. To him, I represented a sort of police force, making impossible demands, but we soon found that we had common objectives for AXE. We both wanted a first-rate product, and we wanted it as soon as possible. I became, I suppose, a sort of moderating influence in the relations between the more fanciful Ericsson engineers and Bengt Gunnar's crew.

Today, we believe the long-drawn-out evaluation process of 1971, 1972 and 1973 was necessary. It caused some conflict. Allegations of unprofessionalism, lying, and deliberate delay were made by both sides, but controversy soon died down. The people at Ellemtel, Ericsson and Televerket had time to get to know what they were really doing, and why we were going about it in the way we were. To the end, there were people on the sidelines at Ericsson who maintained that we had got it wrong, and that AXE would never work. Their doubts persisted until we got the French order, which proved that AXE *would* sell, and that it stood up well to the competition.

Bengt Gunnar and I have become great friends. In September 1975, he had his 50th birthday—a major occasion in Sweden, calling for proper celebration. During a PK-TS lunch, I presented Bengt Gunnar with a scroll, making him an honorary member of the AXE Fan Club. (For another year, he and I were the only members.) The Ellemtel staff gave him an appropriately elaborate party, at his summer cottage on an island outside Stockholm. A few people from Televerket and Ericsson, including me, were invited, but the main body was made up of his friends in Ellemtel. We assembled, about 50 of us, with a large barbecue, fireworks, and a small cannon, to get a boat to the island. As we approached, we sent up rockets and fired the cannon—very dignified, we thought, as the launch to a party. We were inside Swedish naval waters, near the main naval base on the east coast, but maybe the Swedish navy were having supper. It was the area which has, in recent years, been so much in the news internationally, as the playground for Russian submarines.

We never found out what the navy thought. We saw no submarines. We had a hell of a party.

A SECOND LESSON FROM THE GOSPEL ACCORDING TO ST JOHN

Which switch?

By now, of course, it's clear what switching does. It's time to take a look at how it does it.

When we wrote the requirement specification, we emphasised the importance of designing a system which would incorporate digital switches. But we knew that digital switches were still in their infancy, and that components technology had some way to go before the digital switch could become an economically attractive possibility. We had to have a system which was competitive without digital switches, yet would need the minimum modification when digital switches came along.

This had led to the concept of autonomous self-contained subsystems, two of which would provide the switching components of the system—the group switch (group stage); and the subscriber switch. It was clear from the beginning that the first of these to go digital would be the group switch—the technology for the components was nearer to being available.

At this point the cord circuit comes back into the picture. One of the functions of the cord is to feed the direct current to the telephone line. In an analogue system the cord circuits may be arranged in a common pool reached through the group switch. But if the group switch is digital it will not pass a direct current. Foreseeing the early introduction of the digital group switch we situated the cords between the two stages, including them in the subscriber switch subsystem. It may be a wasteful solution in a traffic engineer's eyes, but it did, of course, mean that the system structure was independent of the switch technology chosen for the group switch.

But the first versions of AX would be entirely analogue, and we had to select the particular version of analogue switch technology to be used. It became a major issue.

For a small fraternity of telephone engineers, electromechanical selectors and switches have always had an awful fascination. The Strowger and Rotary systems, and Ericsson's own 500-point selector (the pancake selector) were brought to very high peaks of excellence.

The crossbar switch was slow to gain acceptance, but was a major breakthrough. Its main feature was that it had no sliding contacts. The wear on its contacts was slight, and the use of precious metal as the contact material improved transmission quality. Maintenance costs fell dramatically.

The code switch which Ericsson used was on the crossbar principle, but was simultaneously of higher capacity and more compact. Comparatively speaking, however, it was slow —entirely adequate for an electromechanical system, but not up to the speeds called for by electronic control.

The computer-control systems of the 1960s had led engineers to concentrate on two other types of switches: reed switches and miniswitches. Each had its variations, but here is a brief description of the underlying concepts.

A reed element is a pair of soft iron reeds or tongues hermetically sealed in a glass tube.

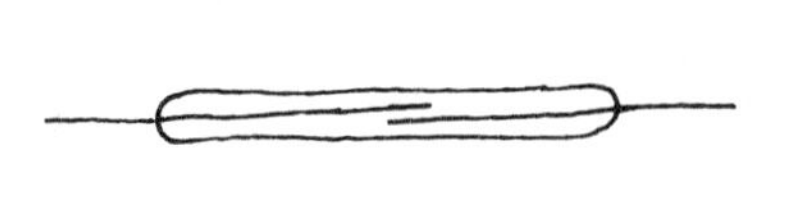

Fig. 17. The reed element

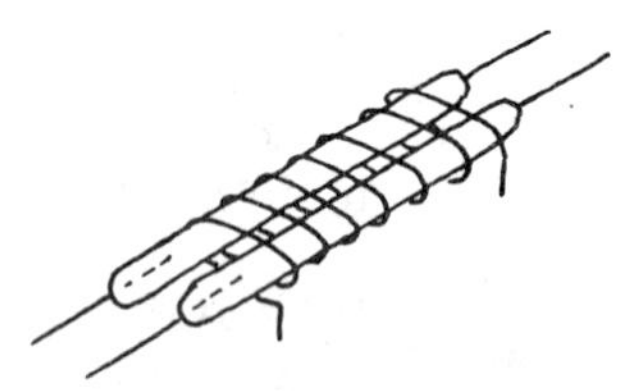

Two-pole reed relay

Inside a magnetic coil (a solenoid) the reeds become magnetised and join together to make contact. Two or more reed elements inside a common coil make a reed relay.

A variation is the ferreed, in which the reed elements are ferromagnetic permanent magnets. The contact is opened and closed by sending pulses through the coil which change the direction of magnetic flow, reversing their polarity. Western Electric used ferreeds extensively.

A switch made up out of reeds is a matrix with a reed at each matrix crosspoint.

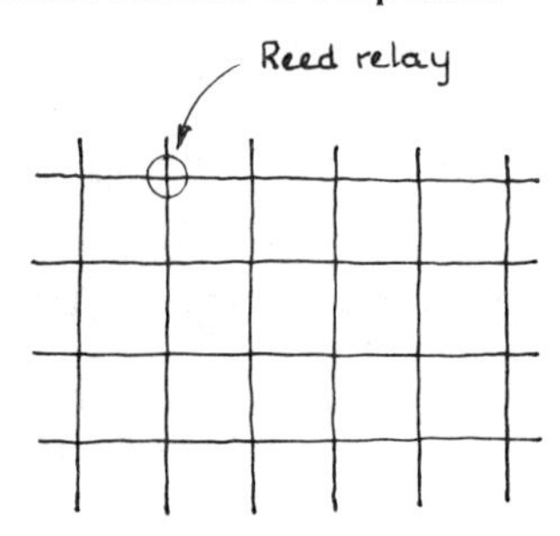

Fig. 18. A 4 x 6 reed relay matrix. One two-pole reed relay in each crosspoint.

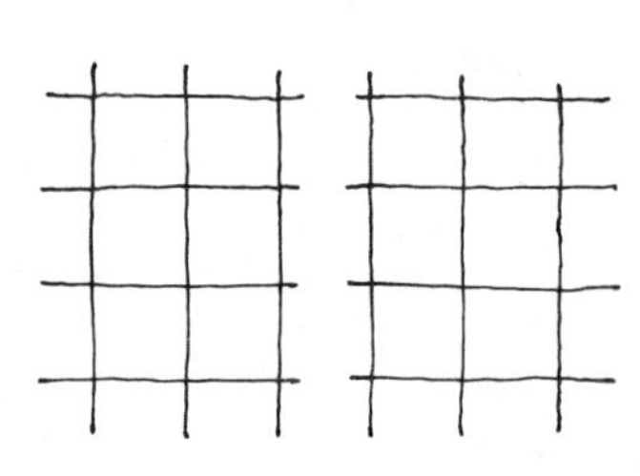

The same matrix cut to make two 4 x 3 matrices.

Reed relays are tiny. They are mounted on printed circuit boards (PCBs), with the circuitry of the matrix etched onto the board. By wiring board to board, we can extend the matrix practically at will, both horizontally and vertically. There are many possible combinations. The art of building a switch—group switch or subscriber switch—is one of selecting combinations to match the traffic capacity required.

Reed switches were in use in ITT's Metaconta 10C and 11A systems; in the Philips PRX system; and various others.

The miniswitch alternative to reeds is also a form of contact matrix, but without relays at the crosspoints. The contacts to be made are indicated by common "bars" or "fingers", and then closed simultaneously by an operation which also locks them in place.

To design a miniswitch within stringent limitations of space and cost is an irresistible challenge to engineers of a given breed. Ericsson and Televerket both had several examples of this species.

There were three existing miniswitch designs to choose from (for the nostalgia record, they were a regular miniswitch; a shunt-field switch; and a bar-operated switch). The reed switch was the fourth possibility.

We had plenty of fuel for argument. And we had plenty of lengthy, heated arguments.

The way the problem was solved is interesting.

Our president was then Björn Lundvall, an engineer who could hold his own in a technical argument. At a large meeting, the different factions presented their cases. The reed switch won.

The main reason was that the manufacture of reed switches closely parallels that of electronics. Reeds can be made and tested on an electronic production line. And, sad though it may be, we knew that the electromechanical era of telephone switching was in its twilight years.

It must be rare, these days, for the president of a large, international concern to settle a fundamental question of technology—and settle it right!

A pause for breath

Now that AX had passed from being a proposal to a development project, we assigned a standard Ericsson acronym to the new product. From now on, we had AXE.

So where have we got to with the AXE revolution?

We've looked at the change-over from manual switching to automatic switching, the first fundamental revolution in telephony, which removed the need for millions of operators to route billions of calls.

We've noted that when the operator moved out, she took her brains with her. And we've seen that this left a gap that had to be filled. We've noticed common control creeping in as a mechanical brain in the form of markers and registers. Then we've seen the electronic brain, stored program control, beginning to come into use. This meant a certain lack of rapport between brain and switches, since they were too slow for the electronics, so we began to see faster switches. Whether electronic or electromechanical (as the reed switch is) they were analogue.

We've also followed the reasoning at Ericsson leading up to the decision to design a completely new system. For this new system the switching would initially be analogue.

In a moment, we shall see how this new system was actually designed, and brought to the point where we had a product we could sell.

But over the whole story so far has loomed the formidable presence of a technology we have hardly discussed, digital switching.

Before we look at the full story of AXE, we must get to the bottom of digital telephony.

The digital revolution

All the switching we have been discussing so far has handled analogue signals—each channel has been carrying a *continuous* signal, varying according to the movement of the diaphragm in the telephone.

The best way to approach digital *switching* is by looking at digital *transmission*—in the form of pulse code modulation (PCM).

PCM is a method of transmitting several telephone conversations simultaneously over a single pair of wires (or, in practice, two pairs—one pair in each direction).

It was invented by Alex Reeves, an Englishman working for ITT in Paris, in 1938, but the components technology to make it a commercial proposition just wasn't available. It only became a going concern with the invention of the transistor, and the development of integrated circuits. It's been in operation for about 20 years, and is particularly popular in city networks for providing additional circuits on existing cables between exchanges.

PCM works on a sampling basis. If you take a signal like the electrical wave produced by speech, and sample it—dip into it—with a frequency at least twice as high as the frequency of the actual signal, you can transmit just those samples and the signal can be reproduced without distortion at the receiving end. This is called the sampling theorem.

Here's what it looks like on a time diagram.

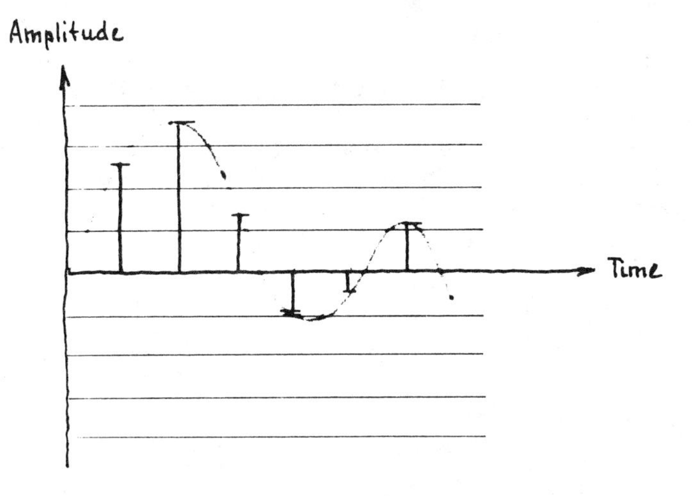

Fig. 19. Sampling

In telephony, speech usually occupies a band width of 300 cycles per second (300 cps) to 3000 cps, so for PCM, 8000 cps has been agreed internationally as the standard sampling frequency.

The point is this. If, instead of transmitting a continuous signal, we're using a line to transmit a pulse, or sample, only at every 125th microsecond, there is room between the samples to send *other* samples, from *other* signals (other conversations) along the same line.

In fact, two international standards have developed. One uses 24 channels per line, the other 32 (though the capacity is normally set at 30, since the other two channels are not used for speech).

After sampling, the second technique is coding, to identify the value (or amplitude) of the signal at the sampling point. To try to send a real pulse of the correct amplitude is far too risky; so many things may happen along the line to distort it and change its value. It's much safer to code the value of each sample before transmission, and to decode it at the receiving end.

The code used is binary. The value of each sample is identified by one "word" of eight bits, each bit offering a position into which we can insert a "1" or a "0", one or zero.

Eight bit positions offers 256 possible combinations of ones and zeros, but we reserve one bit in each word to indicate whether the sample is positive or negative—which halves the possibilities. We are left with 128 possibilities, each of which may be positive or negative.

To see what the value, or amplitude, of the sample transmitted really is, we have only to compare the coded value with an agreed scale, on which it will fall within one of 128 positive or 128 negative areas.

Since there are two standards, there are naturally two different models (A-law and μ-law), but the principles are just the same. Fig. 20 may help.

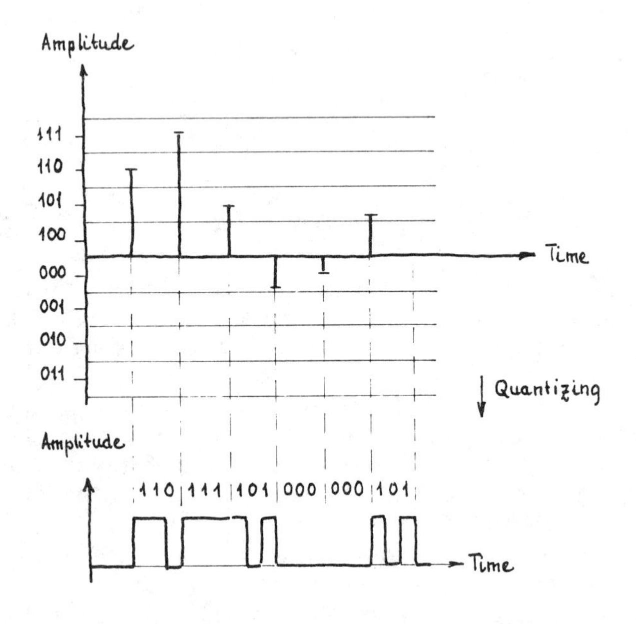

Fig. 20. Principle of quantizing and coding. In this example, we are using a code based on only three elements.

Transmission speed is 64,000 bits/second (64 kb/s), so for one signal we transmit *one* word (stream) of *eight* bits at 64 kb/s every 125 microseconds.

Let's take the 32-channel standard. Between the samples of the original conversation (in channel 2 in figure 21), we insert the other 31 channels—or rather, the samples, in binary code words, of the other 31 channels (on which conversations may or may not be taking place).

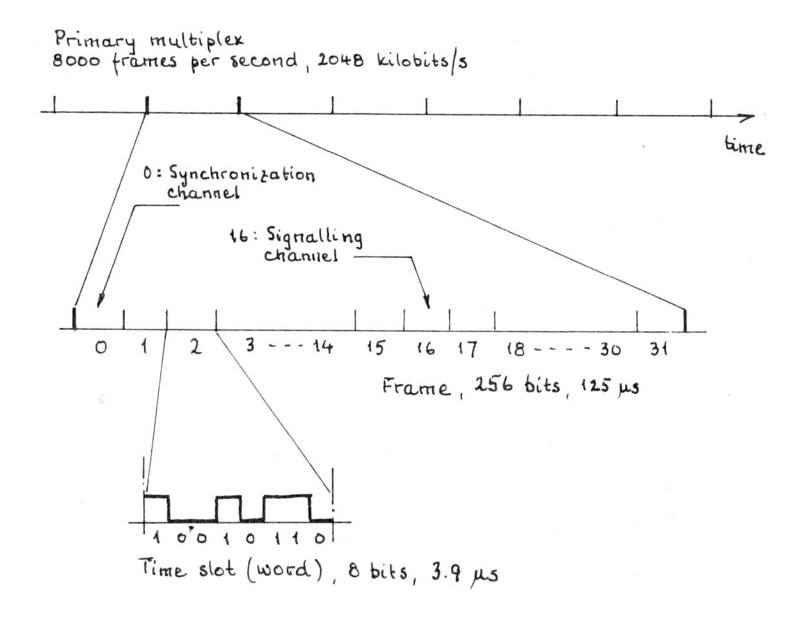

Fig. 21. Primary PCM multiplex: frame structure, 30 PCM channels / plus synchronization and signalling channels.

If you add up the whole lot, you have a stream of bit positions with a full frequency of 32 x 64,000 = 2,048,000 bits per second (usually described as 2.048 megabits per second —2.048 mb/s). You can "multiplex" still further—the standard hierarchy also contains 8 mb, 32 mb, 128 mb and 256 mb systems, where an 8 mb system, for instance, provides 4 x 32 = 128 speech channels, and so on.

(You can imagine old Jeans trying to figure this one out, but if he managed it, you can. Anyway, the figures are not really important for the main story.)

In the late '60s, with PCM widely in use, the principles of digital switching were beginning to arouse interest. PCM is a digital transmission system, which can handle only digital-bit-stream signals. To use it with switches designed to handle analogue signals meant that the signals had to be decoded—transformed from digital back to analogue—before they could be switched, and coding and decoding represents a cost. Consider a group of circuits between two exchanges—both analogue. Basically, there are two ways of providing these circuits—you can use a cable containing the same number of pairs of wires as there are circuits; or you can use PCM transmission, and use just two pairs of wires for each 30 circuits. Wires cost a certain amount per metre, and coders and decoders cost a certain amount each. PCM will be cheaper only if the distance between the exchanges is greater than x. X depends on several factors, but is typically around 8 km-10 km.

When, in the early days of developing AXE, we started to talk about digital switching, we were looking for ways of building switches which could switch digital speech channels. If this were possible, we could eliminate coders and decoders (the combination is called a codec), and the costs of switching and transmission would be reduced. We were beginning to talk about IST, integrated switching and transmission.

An analogue switch sets up a *physical* connection between two points, an inlet and an outlet.

In a digital switch, the incoming circuit or channel is represented by a position in *time*. At a given moment in time, the eight bits arriving on an inlet, or pair of wires, belong to a specific channel—as do the eight bits arriving 125 microseconds later. (In digital switching, 125 microseconds is a long time!)

On the other hand, the eight bits arriving *immediately* after the first eight belong to another channel, and the conversation will probably be directed to a different address.

So digital switching is switching in *time*. We must be able to transfer eight-bit words from one time position to another. The most simple digital switch is illustrated in Fig. 22, on the following page.

The switch consists of a speech memory of 32 memory areas. Each memory area has room for eight bits. On the incoming line, we have a stream of bits arriving at 2.048 megabits per second. This bit stream is divided into 8-bit words. Word 0, word 32, word 64, and so on make up the first channel of the system.

The clock drives the "contacts" shown on the left (which are actually transistor gates). It steps through its 32 positions at a rate of 8000 steps per second. In position 1, it closes the contact at memory cell 1, so that the eight bits of this word enter the memory. In position 2, it loads the second word into memory cell 2. And so on—until the 32 words of the first cycle have been stored, when it goes back and starts again.

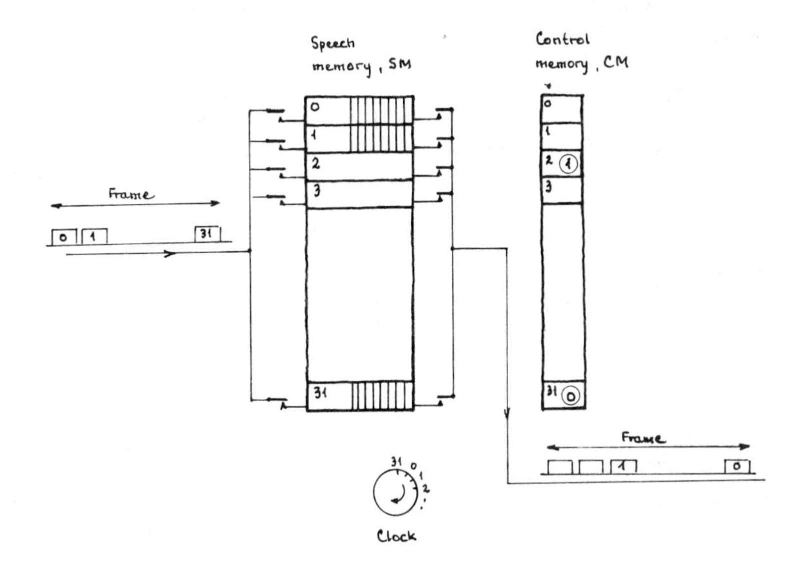

Fig. 22. A time switch. Word positions within the frame on the incoming line appear in other positions on the outgoing line.

For it to load the next cycle, the memory cells must, of course, have been emptied. This is the second process, and it works like this.

On the right in Fig. 22 is the control memory. When the subscriber dials, the control system of the exchange picks up the dialled digits and writes the information of the selected route into the control memory.

In Fig. 22, we have 32 channels on the outgoing line. Let's assume that the control system decides that incoming channel 1 is to be connected to outgoing channel 2—in other words, a "1" is written into CM position 2 (the small figure 1 in the circle). In the same way, incoming channel 0 is connected to outgoing channel 31.

The same clock steps down through the memory cells of the control memory. It stops for a fraction of a second by cell number 2 and analyses its contents. It contains the figure 1, so CM closes the contact on the right-hand side of cell 1 in SM and the eight-bit word passes out on channel 1 of the outgoing line. When the clock reaches CM 31 it finds the number 0, so the contact to the right of SM 0 is closed and its eight-bit word is passed out on the line on channel 32.

In this description, we've used two words perhaps a little loosely. The first is "connect". In fact, what we do in a time switch is store each incoming sample for a moment, and take it out of store a moment later—the length of the delay depending on what out-going channel has been chosen. The word "channel" too is being used loosely; another definition is time slot—the process of switching is actually a matter of changing time slots.

Pulse code modulation is also called "time division," and a switch of this kind is a time switch.

The example we've worked through is on a small scale. A typical size for a time switch block would be 512—16 PCM systems connected each side, each memory with 512 cells.

To build bigger switches—such as group switches (and you know what they are, now!)—it is usual to group together a number of time switches and introduce a space switch. A space switch is a 32 x 32 matrix with electronic crosspoints—gates. The space switch is also controlled by a control memory, which activates the crosspoints for the split second required to pass through the eight-bit word.

There are still people who think a space switch is some kind of analogue switch, but an analogue switch provides a continuous metal-to-metal contact for the duration of the conversation, while a space switch, as part of a digital switch, sets up the path only for the short time necessary to pass the word through.

In the space switch diagram, Fig. 23, each horizontal and each vertical line represents eight wires corresponding to the eight bit positions of the PCM words.

The grid shown is a gate matrix, that is to say that in each crosspoint there is a gate, an electronic relay, which controls eight contacts. The gates are operated under control of the control memory so that, for a given call, which has been assigned an internal time position, the gate is opened (i.e. the eight contacts are closed) momentarily at the same time as the gates in the time switches. Thus the contents of the incoming speech memory are transferred, through the space stage, to the outgoing speech memory and placed in the memory position decided by the control system. When the clock steps to the next position, the control performs the same transfer of memory content for a different call.

Fig. 24, on page 66, shows, on a small scale, the full group switch structure, with its time—space—time stages.

The AXE system uses this arrangement. There are other variations on the digital group switch theme, such as space—time—space and time—time—space—time—time.

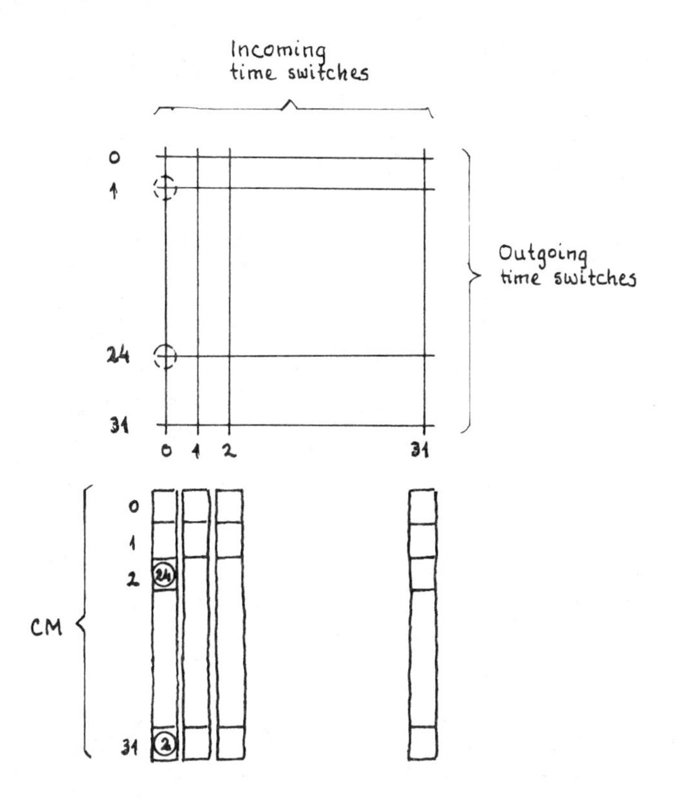

Fig. 23. A space switch. The control memory for each vertical corresponds to the CM of each incoming time switch. In clock position 2, the space switch will connect incoming time switch 0 to outgoing time switch 24, and in clock position 31, incoming time switch 0 to outgoing time switch 2.

As you see from the diagrams, digital switches are to a very great extent built up of memory—speech memory and control memory—and obviously it was the continued reduction of the cost of memory components that made digital switching a workable proposition. In the AXE control system, with its central and regional processors, we also find substantial amounts of memory, and I think this is a good place to make the comment that it was to a great extent this reduction in memory cost that made it possible to implement the modular software of AXE—because the modular concept as applied in AXE does force the designer to use memory. Very obligingly the component industry, largely in the US and Japan, was rapidly developing ever more compact memory chips, in a most timely way. A switch in time . . .

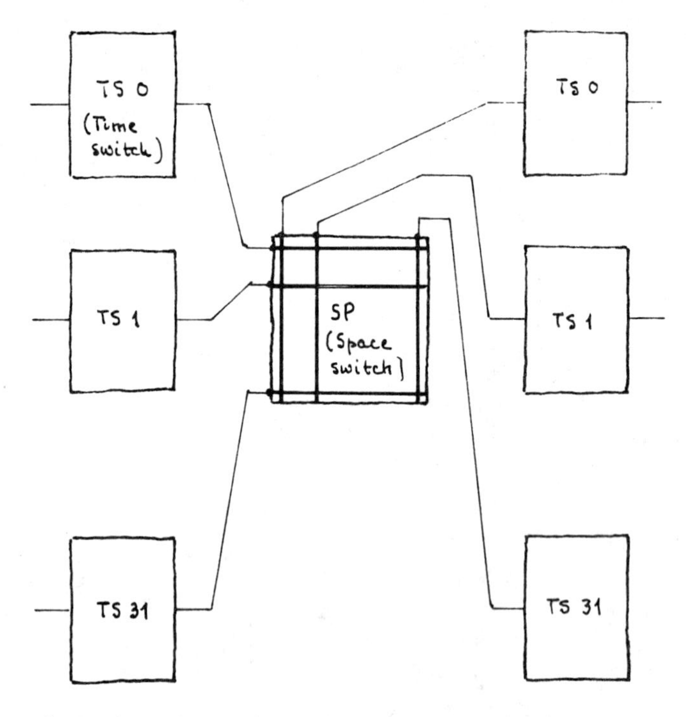

Fig. 24. A time-space-time digital group switch. Each time switch connects 16 x 32-channel PCM systems, a total of 512 speech channels. With the 32 x 32 space switch, SP, shown, the system capacity is 32 x 512=16,384 channels.

Putting it all together: the AXE SPC switching system

You now know as much about telephony as many of the people to whom we eventually found ourselves presenting AXE.

What follows is what we would include in those presentations—though obviously we used to shape the presentations to match our audiences, the time and knowledge they had, and what we knew of their idiosyncrasies.

To make the system easy to handle, we split it up from the beginning into individual modules. We chose as our fundamental design base the principle of functional modularity. And we adopted as our approach a "top-down" analysis of the telephony functions needed to meet the requirements of the network.

In other words, we saw the telephone exchange as just one of the components in a complete network. Instead of being some sort of platonic ideal of an exchange, it was just a bit that had to do what the network required of it.

This really was a new approach. The traditional way of doing things was to design the perfect exchange, and then try to fit it into different networks.

Here's a hierarchy diagram which shows the structure of the system as we perceived it.

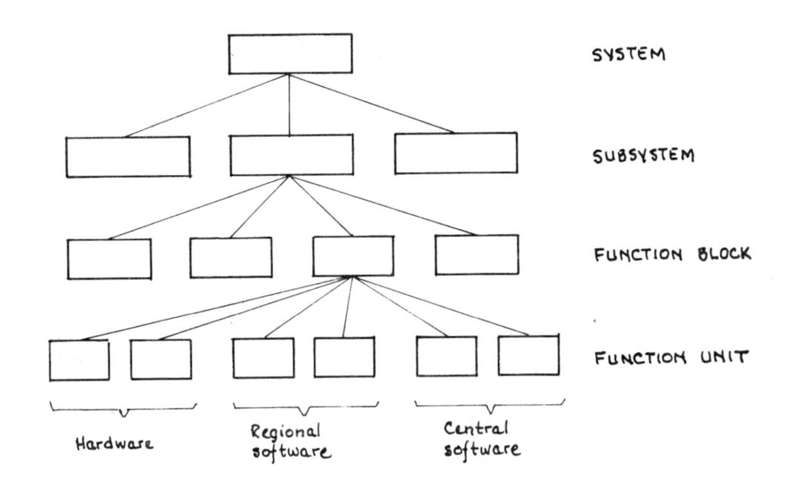

Fig. 25. Functional system hierarchy of AXE.

What's important about this approach, and what we kept trying to demonstrate, was that we were not concerned with *how* a given function would be provided—even at the functional level. We described what a subsystem or a function block had to *do*, in functional terms only, and we left it completely open whether it should be done by hardware or software.

That decision was made from the bottom up, starting with the function units. At that level, if hardware seemed the best bet, we designed hardware, if software, software.

Each box in the hierarchy diagram represents a module, and the point was that each module was autonomous—it was a stand-alone piece of equipment that did not depend on any other module for the way it worked. All that mattered was that its *output* should be able to be handled in one standard way by any part of the network, and that it would respond to *input* of a given type and format.

One example of this principle is worth mentioning specifically. We could have taken one of the processor designs we had knocking about at the time, and decided to use it as the central processor for AXE. Developing a processor is arduous, time consuming, and expensive, and such a proposition would have been attractive. In fact, it was suggested.

But this would have meant that the rest of the system would have had to be designed round one major component. The capabilities of that component would have been a restricting factor on the whole design project. Instead, the central processor was defined and specified as a natural result of the analysis of functions, and became an integrated part of the system—a subsystem within the control system.

Software was designed like the hardware. Previously, software had been arranged in two fields, programmes and data. Any portion of the programmes was free to access any of the data. This meant that a fault in the data which was caused by a faulty piece of programming operating on the data, would spread through the system as other pieces of programme picked it up to use. This is a not uncommon phenomenon, known as data mutilation or data fault proliferation, and it could cause the whole system to malfunction. It was possible to detect such a malfunction, but very difficult to track it to source. It usually meant restarting the machine with clean data, and a long process of fault analysis.

With autonomous modular software, on the other hand, each module has its own programmes and data. Its data can only be accessed by other modules through its own programmes. Signals from module to module are standardised. If a signal from a module fails to conform with the standard, it is detected immediately, and we also know immediately which module caused the faulty signal.

We began to refer to the traditional structure as spaghetti software. The software of AXE was referred to as modular, with software protection.

The modular structure of AXE had other advantages. With autonomous modules, we can modify or replace a module without affecting the rest of the system. As long as a module works with the standard signal interfaces, it is a functional part of the system. As a result, we have been able to introduce new technology into AXE smoothly and naturally. As new memory components have become available, for example, they have been introduced, and have reduced the physical size of some elements within the exchange dramatically.

The structure also meant that we could supply either analogue or digital switchblocks for the subscriber and group switches—provided that their outputs conformed with the

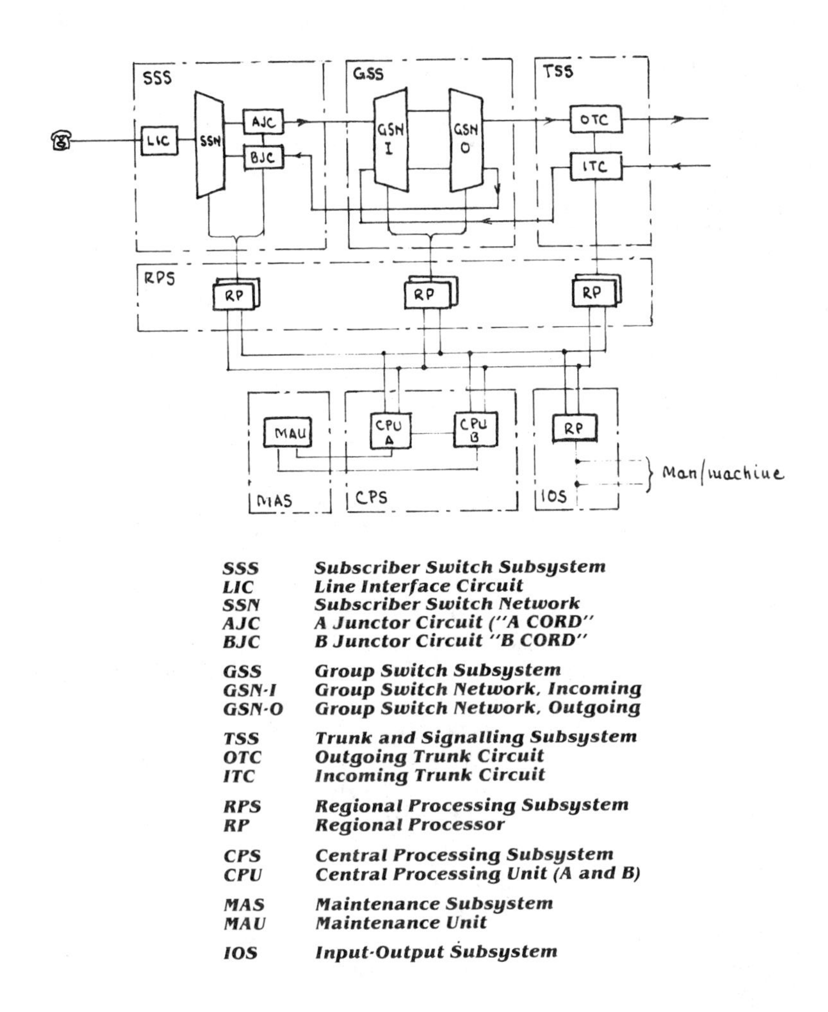

Fig. 26. Hardware structure of the AXE system, equipped with analogue switches in SSS and GSS.

standard for the system. When we reach the marketing story, you will understand how extremely important this feature became.

For the moment, let's confine ourselves to the basic structure of AXE, and spend a few minutes with a diagram, Fig. 26, above, representing the hardware.

The telephony system, APT

The boxes in Fig. 26 are all hardware. In the telephony system, APT, they are circuits and switches. In the control system, APZ, they are the circuits that make up the processors and the associated equipment.

Starting from the left of the APT system, the subscriber lines come in and terminate on a line interface circuit, LIC. There is one LIC for each line. In the first version of AXE, there were four LICs, mounted on a printed circuit board.

The funnel-shaped box which comes next is the subscriber switch network, SSN. In the early version of AXE, this was a three-stage reed relay matrix, which "concentrated" or switched the subscriber lines towards a number of cord circuits.

Note how the traditional names and functions still crop up. Cord circuits are still necessary, but we did give them another name, *junctors*. There are two types: AJC for originating calls; and BJC for terminating calls. Just as it always did, AJC supplies the feeding, DC, current over the lines when the subscriber lifts his phone, and gives him a dialling tone.

When the subscriber dials, a relay in AJC follows the dial pulses, and a contact on the relay is scanned by the control system, which stores each digit.

The Group Switch Subsystem, GSS, comes next. This is in two stages, GSN-I (for incoming) and GSN-O (for outgoing). Each is a three-stage reed matrix arrangement. *All* calls pass through the selector from left to right.

If the call is for a subscriber on the same exchange, the group selector takes it straight back through to the BJC, a terminating cord circuit in the subscriber group indicated by the dialled number. BJC selects the dialled number through SSN, picks up the LIC, and rings the subscriber. When he answers, the ringing stops and BJC is connected through. The call is set up.

If, on the other hand, the call is for a subscriber on another exchange, GSN picks up an outgoing trunk circuit, OTC, and when the exchange at the other end is ready to receive, the control system will direct the appropriate digits to be transmitted.

We can see what happens then by following an incoming call on this exchange. It arrives on an incoming trunk circuit, ITC, and is connected to a code receiver, KM. (Careful—this "code" is not the same as the code in pulse code modulation; it is merely the form in which the digits are received—which could be ordinary dial pulses in a step-by-step system.) KM receives the digits and stores them in the control system as a record. It is then disconnected, and the call is set up through the group selector to a BJC. BJC sends it through the subscriber switch to the line to be called.

The control system, APZ

Control functions in a telephone exchange vary in frequency and in complexity. There are numbers of functions which happen very frequently, but are quite straightforward and do not demand much intelligence.

Scanning the subscriber lines is an example. Scanning must be continuous, since we never know when or which subscriber is going to make a call. Every line, or rather every LIC, is scanned every 5 milliseconds. Scanning is just a check to see whether a subscriber has lifted his handset. From a computer point of view, it's just a matter of registering yes or no, of noting whether there has been a change since the last scan, simple and straightforward.

On the other hand, there are also more complex functions to perform—the analysis of a number called, for example, to determine the routing of a call. Such analysis is not regular, it only occurs once for each call, but it is fairly frequent.

At the most complex end of the scale, we have a set of functions which are brought into play only in cases of trouble. When something goes wrong, a whole set of functions is called into action to determine the nature of the fault, to pinpoint it, to find out why it occurred (diagnosis), and to take action so that the fault will not spread to other parts of the system.

If we draw a diagram plotting the complexity of functions against their frequency, we get a set of points something like this in Fig. 27 on the next page.

When we were structuring the hardware for the control system, we had to decide how to distribute the processing power—where in the structure to locate how much of its intelligence.

The curve you get by joining all the points located on the graph by the functions analysis could be said to define the total processing power required for the system. The larger the exchange, of course, the more processing power is required—but power does not rise proportionally to size.

One solution would be to have one large processor to do all the work. Its size and capability would be equivalent to the complete area of the graph. It would have to handle maximum frequency and maximum complexity. It would, for

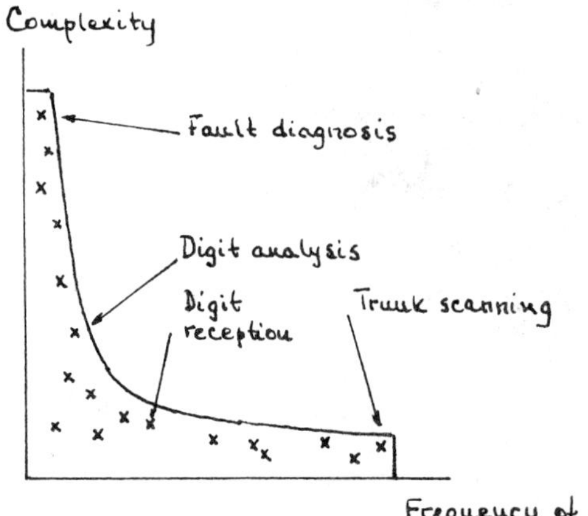

Fig. 27. Relation between complexity and frequency of occurrence of the functions of a telephone switching system. Functional analysis provides a set of discrete points, the envelope of which describes the processing power required for the system.

instance, have to be able to perform the most complex and exceptional functions every five milliseconds. It would be an enormous machine—and enormously wasteful. (All the capacity *above* the curve in the diagram would be wasted.) With this solution, each size of exchange would have its own size of processor—or, even more wasteful, the same big processor would be used for every size of exchange.

At the other end of the scale, we could arrange to have the exchange divided up into subsystems of the maximum size, each module with its own fully capable processor.

In this solution, each processor would be able to handle all the possible functions, regardless of complexity, but the frequency with which it did so would be reduced. The method could be represented by a set of rectangles on the graph.

This is a workable solution in theory. In a big network, it would mean an alarming proliferation of self-contained processors, all of which have to be reprogrammed (which is not very difficult) and tested (which is) if the software is changed. There are, however, strong arguments for distributed control, as this system is called, in some areas of some networks.

For AXE, we adopted a common-sense combination of methods. The simple, repetitive functions—for which soft-

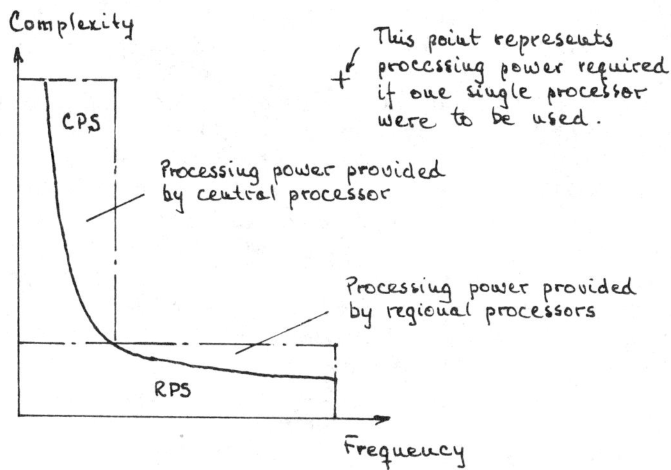

Fig. 28. Distribution of processing power between central and regional processors.

ware is unlikely to change very dramatically or very frequently—are handled by small standardised regional processors (RPs). All the more complex functions are referred to one central computer. In the hardware diagram, Fig. 26, the arrangement is shown with a system of "buses" to link the RPs to the Central Processor System, CPS.

The central processor is duplicated (CPUA, CPUB). The two sides work synchronously in parallel. This means that they perform exactly the same tasks at the same time, but at any given time, only one of them is "executive"—only one of them sends instructions into the RPs. This is to provide security—the workings of the two sides are compared continuously in the Maintenance System (MAS), and if a mismatch occurs we know we have a fault. Immediately, MAS sends both sides into a fault-finding routine. When we know which side is faulty, the other side is switched in to become executive.

RPS is a Regional Processor Subsystem. It consists of a number of duplicated processors, the number depending on the size of the exchange. The pairs operate in "work-sharing mode"—under normal conditions the work is shared between them—if there's a fault in one, the other takes over the full load.

The last box is the Input Output Subsystem, IOS. It has the important task of providing the man/machine interface.

Maintenance staff can have access to the system through different types of terminals, or the system may monitor itself and send the results over data links to a maintenance centre.

We've talked about modularity—the functional modularity of the software and software protection, and we've seen how it's reflected in hardware modularity. Each subsystem module—SSS, TSS, GSS and so on—has a specific size. The SSS can handle a maximum of 2048 subscriber lines. This means that a set of cords is arrived at for each exchange by an analysis of traffic requirements—just as it was in the manual exchange we looked at. The module size for the group switch is 512 inlets and 512 outlets. The TSS—Trunk and Signalling Subsystem—is designed in modules of 256 trunks each.

The modularity of the hardware means that the system can be used for all sizes of exchanges, from one with a couple of thousand subscribers to one with about 40,000 (the number we specified originally—it has been exceeded since).

The structure of the system also gives us full applications modularity—different combinations of subsystems can be used for different types of exchanges in a network. Page 21, fig. 11 shows us a local exchange. A tandem switch uses the group switch only, since there are no subscribers connected. An international transit exchange might require an extra subsystem to connect operators' positions. The system is the same, regardless of the application.

The control system, APZ, is always the same, in every installed AXE exchange, except that the memory capacity may vary to match the volume of APT software. You remember that the applications software (APT programs and data) is stored and executed upon in APZ.

We've mentioned already the concept of technological modularity—the important feature which would allow us to introduce new technology as it became available. In Fig. 11, page 21, the system is equipped with reed relay matrices. It's easy to visualise the same system equipped with digital switches.

The first subsystem to go digital was, of course, the group switch. This merely meant putting in new digital hardware to replace the reed matrices, and the addition of one new item —the clock which is a necessity in all digital systems.

The clock is a pulse generator or oscillator, which uses a crystal with a very stable frequency to control the pace of the digital switch. When a digital network is being built up, it's important that all exchanges work at the same pace. Synchronisation is essential, and is achieved by various interconnections between the clocks in the exchanges.

Failure to synchronise a telephone network results in "slips", which can be heard as small clicks during a conversation. In speech, they're not much of a problem, but the same network is carrying control signals and high-speed digit information. It will also very soon be carrying a significant volume of data traffic, and with data, slips can cause information to be lost or mutilated. Synchronisation is vital, and the role of the clocks is vital, so there are three of them in each exchange. At least one of them is likely to be working properly.

The digital subscriber switch, and the death of the cord circuit

An AXE exchange with a digital group switch and analogue subscriber switch still retains the faithful cord circuits, now called junctors.

But once the group switch is digital, the way is open to separate it from the subscriber stage—to remote the subscriber stage in, for example, a rural application.

The remote subscriber stage, or switch, or concentrator is connected by PCM over digital lines, but until the end of the 1970s was itself still an analogue reed switch.

Analogue technology is not ideal for concentrators, and we worked intensively on the design of a subscriber switch which would make the AXE system completely digital. The problem was largely one of waiting for the right components to become available from the manufacturers. At the beginning of the 1980s, custom-designed circuits came onto the market. At last, we could build *system* specifics into individual circuits, and keep faith with the top-down principle.

Today, we're at an intermediate stage. It's obvious to all telecommunications engineers that the next stage is ISDN, the Integrated Services Digital Network, carrying voice, data and text over a single digital network. CCITT and other international organisations are busy trying to establish the international standards. It's slow work, but at least the telecommunications industry appreciates the benefits of standardization more than the computer industry does. We shall avoid the chaos that has overtaken them.

With the standards of ISDN established, we shall be able to make the subscriber line digital. Until then, we're all designing switches which are digital (or at least electronic) to connect calls along good old analogue telephone lines with traditional telephones at the end of them.

Going into a digital switch from an analogue line means converting analogue speech signals to digital by a coder/

decoder device, or codec. The digital switch does not pass direct current, and feeding current and ringing, for instance, are engineered into the line circuit. The cord circuit, our last link with the nineteenth century, disappears.

In a digital subscriber switch, codec and cord functions and the functions for supervising and testing the line have to be supplied individually for each line (whereas in an analogue system they could be concentrated). Clearly, the line circuits could become very expensive; they must be designed with the utmost care.

With ISDN, these functions will be moved right out to the subscriber's location, the network terminal.

The subscriber switch is a time switch, designed as a time slot interface bus. This means that when a line is calling or to be called, the control system selects a free time slot on the bus, and the line and trunk chosen are both switched to this internal time slot.

This is how a fully digital AXE local system looks.

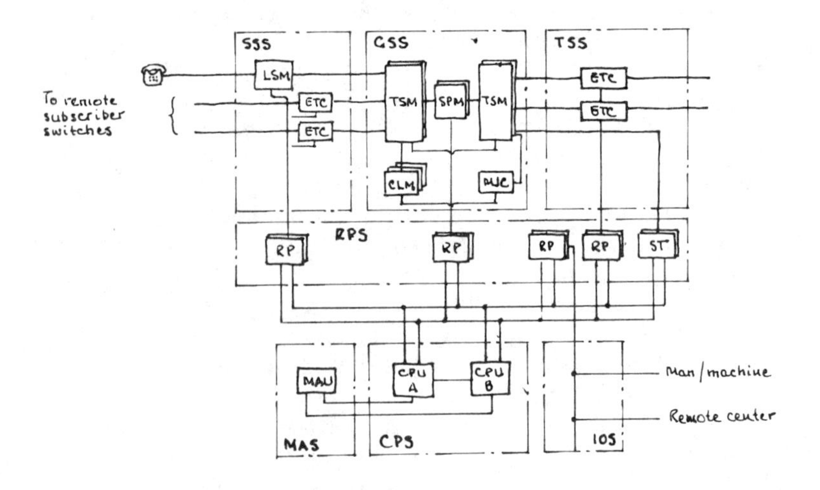

Fig. 29. Hardware block diagram of digital AXE.

Abbreviations:
SSS *Subscriber Switch Subsystem*
LSM *Line and Switch Module*
ETC *Exchange Terminal Circuit*
RPS *Regional Processor Subsystem*
TSS *Trunk and Signalling Subsystem*
GSS *Group Switch Subsystem*
TSM *Time Switch Module*
SPM *Space Switch Module*
CLM *Clock Module*
MJC *Multi Junction Circuit*
RP *Regional Processor*
ST *Signalling Terminal*

CPS** **Central Processor Subsystem
CPU** **Central Processor Unit
MAS** **Maintenance Subsystem
MAU** **Maintenance Unit
IOS** **Input / Output Subsystem

HERE ENDETH THE SECOND LESSON

And the last. There will be no more lessons from the gospel according to St John.

The last section took you into the future. This section leads you back into the past, to the point where the design of AXE was just beginning.

How did we approach it?

By applying the top-down principle, of course!

Each module to be designed was to be autonomous. It was not to depend on any other module for help in completing the particular functions assigned to it. It was to take the input received from another module, perform its own function, then send the result to a following module. Input and result, or output, were to be its interfaces, and should take the form of software signals—data streams, if you like. These signals were to be standardised, with a given format and sequence, so that they could be supervised by straightforward hardware circuitry, alert for the appearance of any rogue signal.

So the first part of the design process was to specify the functions and the interfaces for each module. We called this process "system specification" ("systemation" is a direct translation from the Swedish), and the following Fig. 30 demonstrates a typical piece of system specification. It shows the basic subsystems available in AXE today.

System specification has become a standard discipline, in both senses of the word, and a very important one. No actual design of circuitry took place, no software was written, until the process of specification had been carried right through to the unit level—and tested.

It would take far too much space to reproduce the flow diagrams involved in designing—and testing—right through the hierarchy.

Remember, too, that we were specifying two systems—telephony and data-processing, APT and APZ—along identical lines.

It was a long time before, with the function units defined and specified, we moved on to constructing the hardware prototypes and writing the programs.

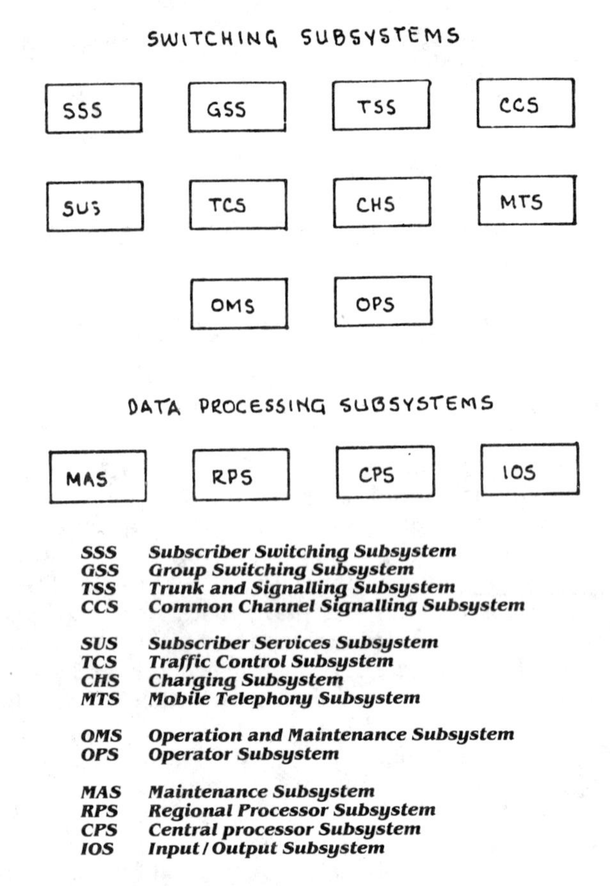

Fig. 30. The AXE system consists of a number of subsystems, each performing a specific role in the exchange. This figure shows all the subsystems, from which exchanges of virtually any type and size can be built up.

Meanwhile, we had made progress with the design of new mechanical housing for the circuit boards and other units. This construction practice was also of a completely new design, based on our experience with the electronics of earlier generations.

Two of the new principles are worth particular mention—a cooling system which cut out the need for forced cooling with fans; and a method of cabling.

To make transport, installation, replacement, and the rest of that side of handling, easy, the requirement specification called for all equipment to be connected by plug and jack. The result was a system with virtually no fixed wiring—that is, no rack cabling. The largest wired unit was a "magazine" which could be varied in size to house up to about 60 circuit boards. The racks were given the form of bookshelves, into which the magazines were slotted, and all the cables—between magazines, to the main frame, for power-distribution, and so on—were plugged in on the front end of the boards.

Each function block occupied one magazine, or in some cases two, while each circuit board represented a function unit.

As for cooling, the racks and housing were so designed that heat escaped through the top and caused an upward draught, sucking cooler air in at the base and up through the magazines. The system works perfectly, even in Saudi Arabia (and Riyadh is *hot*).

An efficient design programme depends to a very large extent on the system design aids, or support systems, available to the designers. The most important aid in the case of AXE is the programming system, which in LMese is called APS. It's a computer-based system, of course, run on IBM or UNIVAC computers, and it includes a programming language, called PLEX (you guessed it—Programming Language for Exchanges). PLEX is a high-level language which allows a programmer to write his instructions in telephony terms—it makes it easier for him to design a system in the way he's used to. PLEX was the first of its kind. It was eventually submitted to CCITT as a proposal for an international standard programming language. Modified and renamed CHILL (CCITT High Level Language) it became just that.

The important point about top-down design is, of course, that every single piece of equipment designed, hardware or software, is designed for the system, and not the other way round. Other companies—and Ericsson, too, in earlier system designs—have tended to select existing designs and force them into a new system, modifying them where necessary to make them fit.

This approach can save time and money, and may produce a system that works. It may even work quite well—though it is unlikely to be the best system possible. Its most serious weakness is that accommodating an existing piece may undercut the requirement specification for the whole which makes for disgruntled marketing men, and short-changed customers.

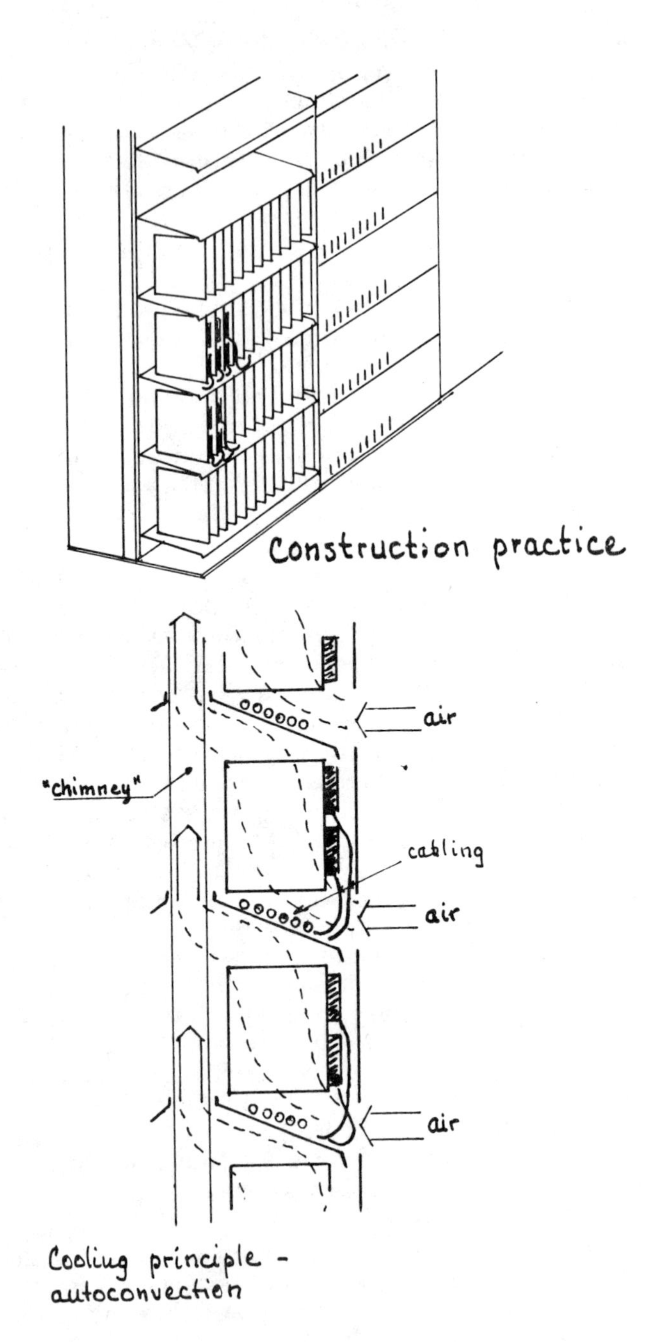

Fig. 31. The mechanics of the AXE system

The progress of AXE development is illustrated by the following figure, which shows the completion rate of the major stages during the years to the cut-over of the first installation—the pilot exchange at Södertälje.

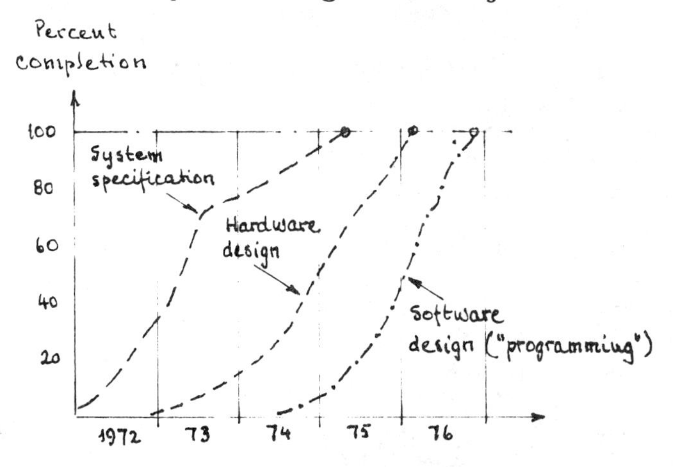

Fig. 32. Completion rates of the design stages of AXE.

I hope that one day Bengt Gunnar and his friends at Ellemtel will find the time and energy to put down on paper the full story of the development of AXE, especially the part covering the years 1970-77. It is of interest not only as a case-history in technology management, but as a story of pioneering, of developing a rather special engineering culture. It is full of human interest. It just takes time to tell.

DECISIONS, DECISIONS

Between the challenge of the design and the thrill of the launch, there are always some niggling decisions to be made with any new product.

What to *call* it, for instance, is often surprisingly difficult. Jeans has known several launch dates rushing up like an express train towards products still known as "Project LX1974" or "that new thing" until a day or so before copy date.

We did indeed have to decide what to call AXE. But we also had to decide where to put the pilot installation, and how to persuade our own people that the product was serious. In our own hearts, we knew that we also had to decide for ourselves whether it actually did work.

Name-calling

So far in this book, the hero has been called AX—and later, AXE.

In 1973, I sent a memo to a selected group of Ericsson people, inviting them to suggest a trade name for the new AXE system. We all thought the name AXE too dull, and we'd been discussing the possibility of finding something with a bit more commercial zip.

We were envious of ITT. We thought that in Metaconta they had a very strong name, and we wanted something like it.

The memo indicated rather vaguely that anyone whose suggestion was adopted would win a worthwhile (though undefined) prize.

Some great engineering minds went to work. These are some of their suggestions.

ELMEX (Electronic LME Exchange)
TELERIC
ERICTEL
ERIMEX (Ericsson Modular Exchange System)
ECCO (Ericsson Common Control Office)
ERCOM (Ericsson Communications System)
COMBEX (Combined Exchange)
ERIPROX (Ericsson Processor Controlled Exchange)
CENTURY (the company's centenary was due in 1976)
ERITEX (Ericsson Telephone Exchange System—a name which has actually been used recently for the Ericsson range of teletex terminals)

and so on.

The clowns of the Switching Division had also been hard at work. Here are some of their proposals.

MESS (Modular Ericsson Switching System)
EUALIA (the official abbreviation for Ellemtel is EUA)
HYSTERIX (the system which copes with overloads)
SEX (Swedish Electronic Switch—we'd expected it, of course, and others in the same vein)
EXOTIX (the system with *all* the facilities)
COMPLEX (the easy-to-understand system)

There were many more of this sort. They were more fun to read, and on the whole I preferred them over the serious ones.

However, the project was shelved for the time being.

In 1976, the question came up again. This time, our advertising manager took the problem under his professional wing.

We were happy with "AXE" for internal use, but we still thought it wasn't commercial. And we still had our ITT phobia—we still liked Metaconta. As a name, that is. AXE was difficult to handle. It has to be pronounced A—X—E (like ARE, AKE, and all our other acronyms). The Finns, who had placed their first order by now, and were following our progress closely, insisted on pronouncing it ax'e.

So G O Douglas got going. He invited a group of us, plus a couple of fellows from an advertising agency (not Jeans's—Jeans hadn't been invented then) to a brainstorming session. We gathered in a congenial setting at ten o'clock one morning, and were instructed to bring forth whatever was in our minds. Nothing happened. Our minds and our large white sheets of paper stayed blank. But lunch acted as a catalyst, and the afternoon was more productive, in fact, disgustingly so. G O's secretary kept a straight face, and dutifully wrote down one unsuitable thing after another.

A few days later, some 30 possibilities were sifted out of this mass of linguistic mayhem, and these were arranged in order of preference by a typically Swedish process of democratic voting.

Five votes went to

NORDEX
MODEX
CROSSTRONIC
ANIARA (a modern space opera)
TELEFUTURA

Six votes went to

CENTURION
CONTACT

And seven votes went to

CENTERIC
ERIC 100
AXE

By now, it was centenary year, which explains CENTERIC and ERIC 100. Personally, I feel CENTERIC had possibilities. It can be pronounced in several languages, including French and English; and in Swedish, it can be transformed slightly into St Erik—the patron saint of Stockholm.

But clearly, the writing, so to speak, was on the wall. After all these exercises, AXE was still cropping up and scoring high points.

In fact, AXE had become a fact of life for us in Ericsson, and we realised that there was no reason why the market should not accept it as just that.

How do you spell that again?

From the marketing point of view, it was of the utmost importance to get a first exchange into service, with subscribers connected, and handling real live traffic, not merely test calls.

To introduce a new product to potential customers there is no argument as valuable as a working exchange. Our presentations may be persuasive and elegant, our documentation excellent, and our lunch perfectly cooked—but nothing beats a reference installation.

The first AXE exchange to be installed was the subject of serious deliberations. It would have to be in Sweden, of course, and Televerket had strong views on location, and how it should fit into the Swedish telephone network. Size, too, had to be determined.

I wrote a memo from Ericsson listing some factors we considered essential. We felt a first exchange, or pilot, installation must be within easy reach from Stockholm, so that we could bring our visitors to have a look at it. We suggested that the premises chosen should be spacious enough for groups of visitors to be taken through. And we very much wanted to have a separate room in the same building which would hold about 20 people, and which we could use for presentations.

When the verdict was handed down we had got almost everything we wanted. Södertälje was to be the site of the first AXE exchange in the world. Sodertalje was to be the name on everybody's lips. Södertälje was to herald the coming of the newborn babe, the third-generation SPC switching system.

But I had forgotten one thing. More than 95 percent of our visitors are from overseas, and nobody from overseas can pronounce, and much less spell, the name of the city of Södertälje. (Indeed, even in this book we've tamed all Swedish spellings, except in this section.)

The first and toughest customer

In the autumn of 1972 we arranged an informal get-together with the heads of the sales departments of the X Division. At this meeting I had been asked to make a brief presentation of the situation of the AX project, and to give a description of the system—such as it was at the time. This was a limited audience; the objective was to inform, and to get some preliminary feedback about what was happening at the different Administrations. What were the plans? How far had the competition advanced? And so on. More important, we wanted to test our ideas for a preliminary marketing plan. It was early days but we felt we needed *something* in the way of a plan, or a model—not yet a strategy, though, that would be putting too grand a name on it.

The presentation I made, the first of very many, was really a retelling in my own words of what I had been learning during my work with the engineers at Ellemtel, plus the early sales arguments that I had been developing (though not the digital switching bit—yet). I remember spending quite a lot of time describing the exact design of the reed switch, with great clarity and enthusiasm. I also remember the comment I got from one of the sales managers, "Surely, you can't be sure that a thing like that works. It seems to me the congestion will be unacceptably high". The name of the questioner shall remain secret—after all we were right at the very beginning of the learning curve.

Later, in 1973, I wrote the first AXE brochure. Strictly for internal use, this was straightforward, simple and cheaply produced—in a cheap red cover. It came to be called "John's little red book". I have been unable to locate one recently. I suppose by now it has been recirculated several times through the paper mills as so much paper rubbish.

You will have gathered that uncertainty about AXE persisted in Ericsson throughout much of the period of development.

Was it really a viable product? Was it better than the other products we had? Wasn't crossbar still selling well?

There were people who felt we should have developed one of the alternatives to AXE. There were even people who thought the competition had better systems to offer.

Sometimes, when I've been asked what was most difficult in the early days of selling AXE, I've answered—a little provocatively—that our most difficult customer was L M Ericsson. The mood of the organisation was such that it's almost true.

Eventually, Hans Flinck, and the few of us on the marketing side who had worked closely with the development and were committed, set about selling AXE internally.

It started with presentations to small groups of sales managers, with discussions afterwards. Then Kjell Sandberg and some other people took a hand. They began to set up training courses which gradually involved most of the sales staff. We ran seminars, two- or three-day affairs away from the office, to build up practical experience in driving home the finer points of the system. We looked at competitors' systems and discussed them, and we involved ourselves in role-playing games, arguing for AXE in front of imaginary customers, with imaginary competitors present and arguing against us.

We started up programmes of specialist staff training, and the production of specialist documentation. We covered installation and testing; engineering to customers' specifications; operation and maintenance; and so on. We produced the first sales brochure for AXE—relying heavily on the design engineers for information, which we then tried to incorporate in words and pictures which would be easier to understand.

We even went as far as arranging training courses, run by English consultants, in how to speak effectively, how to use visual aids, and how to get over speaker's nerves (which I never did learn).

A year or so later, it was clear that all this activity had borne some fruit. AXE was being discussed at coffee-breaks and at lunch; staff were queuing up to get into AXE courses; our companies around the world were asking for training material.

This successful sale to Ericsson was not all due to our activity. The main reason was that we had started to get orders. AXE was emerging from the dreamworld into vivid real life.

IT WORKS

In preparation for the company's 100th birthday one of the many discussion points was how to characterise L M Ericsson. We were active in many fields of telecommunications, but telephone switching was the dominating sector. In switching, we were still living on crossbar and electromechanical technology, but there was a growing feeling that it was becoming important to present a more modern profile. We were spending large resources on development of electronic systems, but we had very little to show in production or out in the field.

The head of public relations wrote a memorandum suggesting a number of activities that might give a new, more accurate, picture of the L M Ericsson of the day. But the most important single contribution would, of course, be a working

AXE exchange in service. This was the demonstration we needed to show that there was something real behind all this talk of electronics, stored program control, and the advantages the new technology offered to the telecommunications community.

At last, late in 1976, the Södertälje exchange was cut into trial service. It was only for three weeks, and only for selected subscribers—all employees of the Swedish Administration living in Södertälje. Getting live traffic on an exchange of a new design is always extremely important. It's the first real test of the system's behaviour under uncontrolled conditions. There were a lot of bugs to be found, and corrected, in hardware and in software. This period of correction, and of intensive testing of the corrections, lasted for a couple of months, and the cut-over date was finally set for March 1, 1977.

The AXE exchange in Södertälje was engineered to replace part of an existing installation of the old 500-point selector system, so the cut-over meant that existing subscribers would be switched from one type of exchange to another. To avoid too much disturbance in the service, the change had to be made at night. A cut-over of this nature always implies a certain interruption of the service when no calls can be made, and when some calls are lost.

Preparations had been going on for some time. They consisted of wiring the new exchange to the main distribution frame, and cross-connecting to the subscriber cables in the old frame, but keeping the new unit blocked by not inserting the fuses. The actual cut-over took place in the main distribution frame room of the building, and consisted of pulling out the fuses on the old side and inserting fuses on the new side. This work began at two o'clock in the morning of March 1, and took some two hours, working in blocks of 20 subscriber lines at a time.

We were following everything closely in the control room. There were a number of alarms as lines that were not insulated well enough were connected, and were detected by the sensitive automatic line-testing. Otherwise, nothing much happened.

We had agreed to put through an early call to Hans Sund. He was due to meet the Director General of Telecom Australia, who had only agreed to visit Stockholm on condition that AXE was actually in service, and Hans was, of course, eager to know that all had gone well. But Hans has a quick temper at times, and it required an act of courage to call him at four o'clock in the morning! As it turned out, this was the first live call through the exchange. It went straight through and rang

Hans at his house in Stockholm. We had it monitored and traced, and the printout shows that it was made from subscriber number 32659 in Södertälje, which was using position 333 in the subscriber multiple. The trace identifies cord circuit AJ number 21, and the group switch positions and link used to reach an outgoing trunk circuit, which in turn accessed the trunk to Stockholm. Some of us signed our names on this piece of evidence. (See the back cover of the dust jacket of this book.)

Hans was awakened at an ungodly hour, but replied with cheerful dignity. He had good reason to be happy. He had, after all, been instrumental in making the difficult decision, and now at last he knew that the thing worked.

This was a great moment for so many of us, but it had been singularly undramatic. Apart from the traumatic experience of getting up at midnight (in my case) any drama had to have some other source, so Bengt Gunnar had brought in a case of bubbly. At four in the morning, we were sitting around having sandwiches and champagne. Then we went to our respective offices.

There were some further problems with the Södertälje exchange, but though they came as a surprise they were solved within a couple of days. Some subscribers complained that they were getting a continuous dialling tone, but we soon discovered that this was because they were used to an old system, in which it's a couple of seconds before you get the tone. With AXE, it's there before the handset has reached your ear.

CHAPTER 3: MAKING IT IN THE MARKET PLACE

LAUNCHING THE NEW SYSTEM

How to buy a telephone exchange

You don't exactly sell a telephone exchange by sticking your foot in the door of a recalcitrant chief engineer. Nor can you sell it through a mail order catalogue. In fact, as you'll have gathered from the story so far, it's not very easy to sell exchanges at all in today's climate of economic depression and cut-throat competition.

In this business, the relationship between the customer and the vendor is often described as like a marriage; once an Administration has made a system choice, the relationship is expected to continue for several years.

This expectation has become even stronger as telephone exchanges have increased in complexity. As complexity increases, so does the Administration's investment in training staff in all the different skills of handling the system—operation and maintenance, planning, management, and so on. Documentation on paper, on micro-film, or on computer disc or tape is voluminous; it takes both space and skill to handle it. Training demands equipment to train on—indeed, many Administrations set up their own captive exchanges. There's a need for spare parts as replacements if hardware fails.

It all costs money, and it all takes time for an Administration to absorb. As a result, an Administration is in no hurry to introduce a new system—as long as its existing supplier behaves himself, gives good service, delivers on time, and keeps his prices reasonable.

Pricing brings in the controversial question of financing. Some international financing bodies—the World Bank, for example—stipulate that their clients must always accept the lowest bid. As a result, some Administrations have ended up with a wide variety of switching systems, even within one city—even, indeed, within one building. You can imagine the cost and the confusion for the Administration in running such a network, and the duplication or triplication of spare parts, documentation and training. Technically, the exchanges will be compatible. Administratively, the network will be a mess. It seems to me that competitive supply can be arranged more sensibly by such techniques as long-term contracts with price guarantees.

Nevertheless, in the last half of the '70s and the beginning of the '80s, many Administrations have gone through the process of selecting new systems and new suppliers. It's a long job, typically involving the following stages.

First, the Administration, or a consultant, writes a requirement specification. It sets down all the characteristics of the new system it wants to incorporate in the network. Today, most Administrations are specifying digital technology and looking for equipment that will eventually evolve, naturally and smoothly, into the Integrated Services Digital Network—they are planning the telephone networks which will very soon be carrying data as well as voice.

The specification will also cover the subscriber facilities to be offered, and the administrative, operational and maintenance features required. Traffic-carrying capacities are important, and so are the different exchange applications in the network for which the system will be used—city exchanges, tandems and transits, international, rural. The Administration may intend to incorporate mobile telephony on a national scale, and so on.

In many countries, local production is a prime requirement—to add to an existing industry, or as the beginning of a new national resource.

The document will probably cover financing—what terms can the manufacturer offer? (In many markets, unfortunately, soft money has become more important than the design of the equipment or its true price.)

The requirement specification is the basis of the tender. The Administration may issue it selectively, inviting certain candidate suppliers to bid. Or it may set up a free-for-all battle.

Candidate suppliers respond with proposals, and preparing a proposal is a lot of work. It involves technical engineering, dimensioning and traffic calculations, specification of material, pricing, and the assembly of comprehensive documentation.

On a given date, the proposals are handed in, and the customer begins the process of evaluation. This, too, can be an arduous and lengthy business, and it's made more difficult by the fact that many of the criteria are intangible, and not easy to compare in strict monetary terms—quality, for example, and service.

This is a time of great tension for the tendering companies, especially where the work of the Administration is performed "in camera", and nobody don't know nothin'. Eventually, there's a decision and the winner is announced, though there may still be a considerable time lapse before the legal contract is finally signed.

The decisive factors vary from case to case. Basic price is always a criterion, though often qualified by financing, and by an evaluation of the costs of operation, maintenance and spare parts. Quality of software and support is as important as quality of hardware, though difficult to assess and overlooked in many cases.

Above all, in an environment and at a time where so many new systems are on offer, a customer will be eager for reassurance that a system is actually proven in operation. There have been many cases in recent history where a customer, seduced by an attractive design or an attractive price, has chosen an unproven system, and then run into difficulties.

The usual problem is delay—it takes the manufacturer longer than he estimated to implement his design and go into production. Even at Ericsson, we've had our problems, but luckily (for AXE) they arose during our learning period with AKE, and our delivery record with AXE is exceptional.

But just occasionally, an unproven system may turn out to be unprovable. Quite simply, it doesn't work. There've been a few dead ducks in the history of telephone switching, and a dead-duck delivery means a near-dead manufacturer. It takes years and years for him to recover, if he ever does.

Tactics and time

In all this, one thing is quite clear; you can't buy a product if you don't know it exists.

At Ericsson, we had to work out how much, how soon, and to whom it made sense to talk about AXE.

For any marketing organisation, the period when a revolutionary new product is under development is frustrating. Income has still to be generated, which means that existing products must continue to be sold. Indeed, in a high-technology environment, existing products may even have to undergo short-term development and refinement.

There is a great danger of loss of impetus while the new product is being developed. It is often best to play down the work that is going on, or to keep it secret, so that sales engineers can continue to sell existing products with confidence.

For two years, the doors around the AX project were virtually closed. At L M Ericsson, only a few small groups were in touch with the work going on.

To the world, we were offering two systems in the early '70s—the AKE system, and the AR family. AKE 13 was a full SPC system, selling well, but only for large national and international trunk exchanges—a market which was naturally limited. The AR family was our volume product. It consisted of ARE 11 (for local exchanges) and ARE 13 (for tandem and long-distance)—both tried-and-tested crossbar systems with computer control. Without SPC, they were sold as ARF and ARM respectively.

Both ARE and AKE were good systems, but both were slow, compared with some of the competition, and AKE had high handling costs. They had an image of old-fashioned solidity, with some electronic make-up applied. But they were what we had to offer.

The world, of course, was still not convinced about SPC, and was still buying bread-and-butter crossbar from us. But we knew that this offered us only a limited future.

Just how limited that future was, even we didn't realise.

Before committing ourselves to AXE, we had made an attempt at technological forecasting. Now, technological forecasting has become a science in its own right. By studying historical technical replacement processes, like the change from sail to steam in ships, or the replacement of propeller engines by jet engines in aircraft, technological forecasters are able (or so they claim) to predict the rates of change in new areas. Their predictions can be expressed as learning curves, or S curves, describing the penetration over time of a new technology, or even of something more intangible like a new management theory.

I made my own first attempts at technological forecasting in mid 1971. The approach was simple; based on the development of our sales of switching equipment over a series of years, I made projections assuming that existing trends would hold as far as the number of telephones in the world was concerned. In the total sales projected, I then inserted a "model" of AXE lines to be installed per year over the next ten-year period. I insisted on using the word "model" as I felt my methods were far from scientific. The model, or scenario as such a projection would probably be called today, also included what I called disturbances in the form of reduced sales during certain periods due to crises of different types—economic, political, or because we lost an important market. I wish I could claim now that I foresaw the oil crisis of 1973. Such a claim would not be very honest, but I did inject my disturbances, and one of them happened to coincide with the oil crisis.

In diagrammatic form the result looked like this:

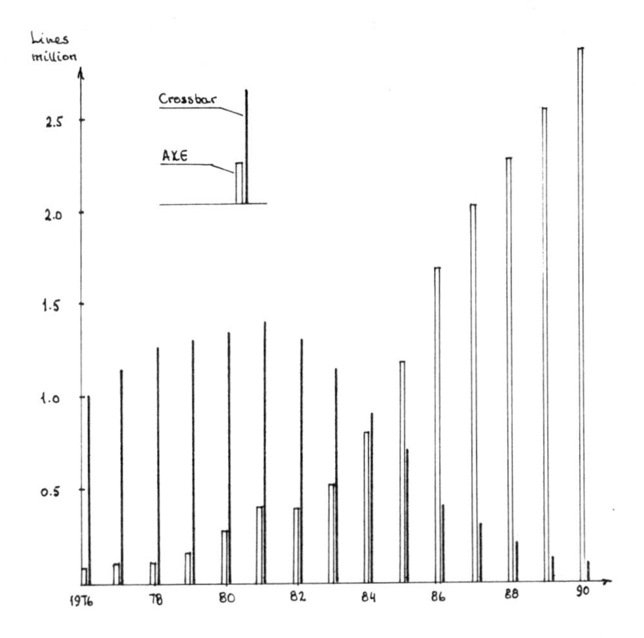

Fig. 33. 1971 model. AXE and crossbar lines installed per year, local exchanges.

The figure shows the projected total sales, and the distribution between AXE and our electromechanical crossbar system. In time these projections became increasingly important in our discussions with our customers and for our planning generally, but 1971 was still early days and the model did not raise much interest. There is, however, one point which is worth noting. The diagram foresaw that crossbar would begin to phase out in the early 1980s, and that we would reach a crossover point in 1984 with equal sales of crossbar and AXE. The numbers I projected were off the mark, but the cross-over point I got right. Quite apart from demonstrating my skills in studying the crystal ball, this shows the resilience of an established system. With a large installed base, a manufacturer will go on delivering the same equipment for many years.

By 1973 we were beginning to get our digital act together, and forming our ideas about the speed of takeover by digital switching technology. In the following diagram (Fig. 34), from 1973, we tried to paint a world picture of this process. We thought at the time that crossbar and other electromechanical systems would dominate the scene well into the second half of the '80s, but more important, we believed then that digital switching would develop at a slower rate than it actually has. Even as I write in 1984, it is, I find, difficult to get exact statistics, so we don't actually know what the situation is today. Here is the 1973 attempt.

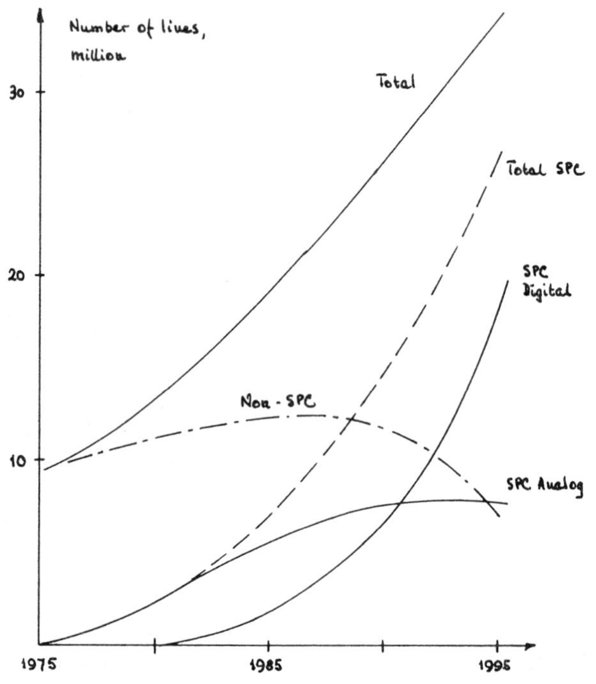

Fig. 34. 1973 model. Number of local exchange lines installed per year—world.

By 1975 we were beginning to get orders and the Ericsson organisation was preparing to handle an "active" product. The design departments had to train staff for designing market applications, the installation department needed to train people to install the exchanges, and the sales departments expanded their training programmes to widen the marketing effort. And, also most important, the factories in which AXE was to be produced had to start installing production equipment and test gear as well as training personnel. They all needed to know for what quantities of equipment they should dimension their efforts; how many people, how many machines? So at this stage the models I had been playing around with had to become forecasts on which to base our staffing and factory space and ordering of components.

I produced the following forecast at this time. It shows number of installed lines in local echanges per year as before, but also attempts to detail the number of exchanges and the number of markets.

	Number of lines	Number of new exchanges	Number of new markets
1976	3 000	1	1
1977	8 000	2	2
1978	30 000	3	3
1979	120 000	15	3
1980	280 000	35	6
1981	480 000	53	5
1982	730 000	80	3
1983	1 060 000	115	4
1984	1 450 000	120	2
1985	1 850 000	160	1

Summing up the numbers, we believed at that time that we would have AXE local exchanges working in 30 markets, remembering that in this business a market equals a country (the exceptions are Finland, Colombia and the USA, countries which have large numbers of telephone Administrations). Further, we thought there would be a total of 584 AXE exchanges working by 1985, and a total of 6.1 million lines in service. We have not yet reached 1985 so it is too early to verfiy how well we were able to forecast—but I feel pleased about the outcome so far—for year-end 1983 I was very close to the mark.

At the end of 1973, the international oil crisis hit the world. In the telephone industry, its effects began to be felt towards the end of 1974. It had a particular significance for AXE.

As the world economy declined, telephone Administrations found themselves short of funds to invest in new plants, or extensions or replacements. This meant that they postponed decisions to adopt any new system as a standard, particularly for local exchanges. What little they did spend on building up their networks, they spent on existing systems.

There were two important outcomes. First, with every month that passed more and more Administrations became more and more familiar with the concepts of digital switching. And second, with every month that passed the cut-over of our first exchange, scheduled for late 1976, came closer.

And we needed that exchange to be able to offer anything resembling a proven system.

Nevertheless, our day-to-day business was hard-hit. Crossbar sales began to fall off especially dramatically during 1976 and 1977. In fact, by 1977 we at last began to realise that the heyday of crossbar was over.

Against this background, we began to formulate a marketing strategy with two main thrusts. We would *inform* all our major customers of our work on AXE, but we would position it as our product of the future, designed to team up with ARE and AKE and provide a complete product range. Meanwhile we would try for an early order for *one* AXE installation as soon as possible in each of our major markets—a sort of field trial and evaluation exchange. This we called the canine concept of marketing control—like a dog peeing on every bush and tree, we, too, would mark our territory.

It didn't work, of course. As a strategy, it failed to take account of a major change in the way the market was to operate. Competitive pressure and the world's economic collapse swept away the very concept of cosily-controlled markets, into each of which one could slip an AXE exchange without fuss. With the larger Administrations taking the lead, project after project began to go through the process of public tender, leading to system choice. AXE grew up in the fierce glare of the public-tender spotlight.

Switching on the spotlight

But in the first half of the '70s, these changes were still to come. We felt confident about the canine concept.

Until the end of 1973, AXE was a well-kept secret. We were both visible and vigorous in the marketing of ARE and AKE. In January 1974, the Australian Post Office placed its first orders for two ARE exchanges, and later decided to up-date the whole crossbar network to ARE by replacing the registers. With AKE, we also managed to secure a number of orders for large trunk exchanges, especially for large international switching centres.

In Iran, admittedly, we failed, in the only tender competition of any size I can recall during this period. It was a large international competition for computer-controlled exchanges, for which we offered ARE. We put a lot of effort into the project, but the contract went to GTE (General Telephone and Electronics, US).

In general, Administrations were lying low. And with AXE, so were we.

But at the very end of 1973, we started to put the canine

concept into operation. The opportunity came when we learned that the Helsinki Telephone Company in Finland was writing a specification, and would be inviting proposals.

This suited us, though we would have preferred to wait another year. It was an opportunity in the Nordic area, on our home ground. If we were to offer AXE, we should naturally have to inform the other Finnish Administrations. (In Finland, the state is in charge of the long-distance service, but most local service is handled by local Administrations.) We should also need to put Denmark and Norway into the picture.

In December 1973, we made our first AXE presentation in Helsinki. In January 1974, we began to widen our net, and I made my first trip as one of a small team presenting the AXE systems. That was to Oslo, and in March I also made two trips to Finland.

In April, we went to Mexico, the country where I was still personally most at home. Teléfonos de México had a large AKE trunk exchange in service already, with others on order, but although it was now functioning well, the company had suffered from our development delays with AKE, and had not forgotten. They were guarded in their response to AXE. "Surely, not another system already!" But our message did rub off, especially on some of the younger engineers.

In these early days, we tended to present AXE as a demonstration of the direction technology was taking—an excuse, if you like, to discuss the requirements that an operating company might impose on a new switching system.

By the time we reached Mexico, we were cautiously beginning to present AXE as a product that might be available—soon. When we arrived in Australia, in November 1974, we knew we had to be prepared to tender with AXE.

We made, in all, some 20 presentations of AXE in 1974. AXE was still an Ellemtel product, and had not been handed over to L M Ericsson. The number of us who knew enough about it to present it was still small, so our teams usually included one or two people from Ellemtel.

But we were getting valuable feedback, and we were developing a convincing presentation sequence. One member of the team would open with an over-all description of the system and its main features. Somebody else would follow with the technicalities of the hardware, and the principles of the software—crucially important. A third speaker might pick up, say, the operation and maintenance features of AXE—which, again, we wanted to stress. We pooled presentation material and became a fairly slick team.

Only once in 1974 did we give AXE any sort of public

exposure, and then in a very low-key way. In September, the International Switching Symposium, ISS, was held in Munich. We had submitted two papers, one on ARE, written by Björn Svedberg and me, and one on AXE. The ARE paper was printed in the proceedings, but not accepted for oral presentation. The AXE paper, written by Kjell Sorme of Ellemtel and Inge Jonsson of L M Ericsson, was presented orally by Kjell.

ISS is held every three years, and is a professional forum, unlike so many other telecommunications conferences and exhibitions today. Fine—we had no desire at all to start the commercial hard sell. We also had no desire to give away too much to the competition; the paper was presented as a discussion of the software handling requirements of a modern system and of a model system which would meet them. In other words, we did present the basic AXE principle of software modularity within a system structure but obliquely.

There were few comments—from competitors or customers. But psychologically, it helped us to feel that we were committed.

Focussing the spotlight

And so 1974 and 1975 went by. We were doing our best to put the canine concept into practise, but with the exception of France, Turku, in Finland, and Australia (and we shall come to them shortly), the customers were coy. Shattered by the oil crisis, they were reducing their purchases of crossbar, and they weren't falling over themselves to put in trial AXE exchanges. We were facing the very difficult decision to cut our work-force—beginning in our own Swedish factories.

"What a hellish year!", said our president, Bjorn Lundvall, in December 1976. And we knew what he meant. He was speaking at a dinner in Stockholm to that year's bronze medalists for 25 years' service. I was one of them.

Nevertheless, there had been some progress. AXE was beginning to be talked about, and to become known around the Administrations and the competition.

In fact, we had chosen 1976 as the year to start our marketing drive, the year for AXE to go public.

1976 was the company's centenary year. Squeeze or no, we would celebrate in style. In the week between May 11th and May 15th, we laid on banquets, receptions, a seminar on telecommunications, and visits to some of our facilities in the Stockholm area. We invited guests from our customers around the world, from the International Telecommunications Union, and from our industry. We had important speakers from every continent.

A problem, of course, was that like other companies in the business, we were in a state of technological transition. We'd spent a hundred years with electromechanical technology, and the bulk of our production was still of crossbar.

True, we had also been working with electronic equipment for many years. Transmission equipment, defense electronics such as radar, microwave, satellite electronics, components. For the centenary, we even produced a fully electronic, one-piece telephone—every guest received one, gold-plated.

But our largest product area was the telephone exchange business. This was what we were famous for, this was where we had made our most important contributions to the art. The centenary celebrations were the natural opportunity for us to do a large-scale launch of AXE.

Several years earlier, we had started work on a three-volume company history to appear in our centenary year. The first two volumes were corporate history, the third was a technical history from the earliest days up through AKE.

For AXE, we needed something more.

We'd been through the French and Australian proposals the year before, and we had quite a lot of material—of a sort. We had overhead transparencies, brochures, technical reports. But they were both very specific and very technical, and the text was wooden, often written directly by Swedish engineers (who frequently have difficulty in writing even readable Swedish). The overheads, mainly drawings and diagrams, were downright dull. But the story all this material was telling was interesting and important. It was a matter of making it acceptable to a wider audience.

Four times a year, the company publishes a magazine, *Ericsson Review*. It's a professional publication, largely devoted to new systems and products, and distributed in Swedish, English, French and Spanish, mainly to the staff of telephone Administrations. We got permission to produce a special centenary edition—in colour. (The company was spending so much on the centenary celebrations, we reckoned nobody would mind spending a little extra, even if it wasn't strictly related to the centenary. *Ericsson Review* has appeared in colour ever since!) The whole of the centenary edition was devoted to AXE—a full presentation of the system, including articles to explain the modular structure of the software.

We even got away with some advertising—not always very popular with Ericsson's management. Again, the mood of the year helped. We had a slight structural problem, since we were still marketing ARE and AKE, as well as AXE. It was felt that all these systems should receive equal exposure. In the end, we ran three advertisements, directed at telecommunications engineers and presenting the three systems under a common banner—*The Switching Expert's Guide to Stored Program Control.* They ran in the specialist press, and I've no way of knowing how much impact they had. Certainly, they helped to put AXE on the map, and since they were pretty technical they found wide use as reprints on many other occasions. We also produced three booklets, each describing one of the three SPC systems.

But none of this was a substitute for *showing* AXE—and AXE was still not in service. The pilot exchange at Södertälje was installed, and being tested; cut-over was still more than 6 months away. Still, we decided to show it.

There's a logistics problem in loading up, say, 400 people, transporting them to an exchange, and taking them through a room where ten people would be a crowd, so we resorted to television. One of the highlights of the entire programme was a 25-minute review by Björn Svedberg of the Group's products, and technical achievements. During this talk, the Södertälje exchange was brought in over closed-circuit TV for everyone to see. In their great wisdom, management had chosen me as presenter, so Svedberg and I had a brief dialogue as I took the audience through the installation. At least the AXE exchange came over very well!

We made a film for the centenary, *Linking People Together,* which had its world premiere during this hectic week. We even made a film of the festivities themselves. It's never been given an official name, and is not shown to outsiders. Privately, we call it *Drinking People Together.*

Performing in the spotlight

Elaborate as it was, the Södertälje TV presentation in our centenary year was only one of scores of AXE presentations we were now giving.

In fact, looking back on the battle, it's clear to me that though our decisive warhead was AXE, with its unmatched range of features and benefits, the method of aiming, sighting and delivering the warhead was very nearly as important.

That method took the form of presentations.

Australia was where these exercises took shape. There, in November 1974, we made our first full-dress presentation. We were to make many, many more. Some were on our home ground in Stockholm. Some were in our customers' own countries, or at conferences and exhibitions.

I was not alone, of course. From the early days, Kjell Sandberg became a regular roving ambassador. Sometimes we presented together, sometimes separately, at opposite ends of the world. Other people—including our chief executive—became honorary members of the team. We prepared a certain amount of material together, but we also tried to work out individual approaches which suited our different personalities. We became a little competitive with each other—good, no doubt, for our own souls and for AXE.

In any presentation, good audio-visual support is important. But it's equally important to know the subject well, and to be interested in it, to speak with enthusiasm and conviction, and to answer questions straightforwardly. Maybe most important of all is to be thoroughly briefed—on your audience, its own view of what's important, and the general business background. Time spent in reconnaissance is never wasted.

One comes to expect the worst from every occasion—the lamp in the slide projector blows, somebody in the audience feels bloody-minded and puts embarrassing questions (planted, perhaps by the competition), there's no chalk or felt pen, the hall is booked for another group an hour earlier than you expected, you develop a serious coughing fit, or the president of the company you're addressing goes to sleep. If all this scares you, you should be in a different business. And anyway, statistically all these calamities don't happen during one and the same talk. (Except, of course, in . . .)

Curiously, presenting becomes addictive!

Exhibitions never become addictive—or at any rate, not to me—but in 1975, we even allowed AXE to put in its first public appearance at a show. *Something* physical is better than even the best slide show.

We chose the Telecom 75 exhibition, held in Geneva in September and sponsored by the ITU. It was considered an important and fairly professional event. We had only a small stand, and we were exhibiting a wide variety of products—transmission systems, cable, an electronic intercom, and so on. We showed just one bay of AXE hardware, with a selection of printed circuit boards. We also had the AXE brochure, but even now we were distributing hard information fairly sparingly. The construction practice of AXE—the mechanical design of the racks; the units holding the printed circuit boards (called magazines); and the printed circuit boards themselves—was new, and aroused a lot of interest.

Inevitably, most of the questions came from the competition. One develops a certain mode of answering these questions—very polite, very guarded. There was the usual contingent of Japanese festooned with cameras. After some discussion, we agreed to let them photograph, though we excluded some boards we felt were a bit too special or which didn't look sufficiently professional, but we were disappointed when they didn't reciprocate. At the Nippon Electric stand, we were told that taking photographs was not permitted by company policy. Funnily enough, company policy had changed when we took off our Ericsson badges and posed as customers. We got our photographs.

Of course, getting this sort of information is a sport rather than technical espionage. You may get some feeling for how advanced in development a competitor's product really is, but the possibility of picking up real knowledge, which one can apply in one's own design or copy to advantage, is very remote. And anyway it's not the details of a system which define its excellence, but its structure—and especially its software structure.

One problem with exhibitions is that they naturally tend to be hardware-dominated—and so many of us fall for hardware. In brochures, in system descriptions, in trade shows, it's the hardware that catches the eye. There is the equipment in all its glory, surrounded by terminals of different sorts. We are proudly shown the printed circuit boards, and told of the state-of-the-art technology employed. The implication is that state-of-the-art hardware produces a state-of-the-art system.

Of course, it does nothing of the sort. Hardware is really secondary—what matters is the system architecture. A soundly-conceived architecture is what guarantees long life to a system, and its use in all the many applications, present and future, where it is expected to serve.

A system is a combination of components, subsystems or other parts, which as a whole will provide a desired set of functions. The system architecture establishes how the work of the total is distributed among the components, and how the components are to work together and communicate with each other. It does this in such a way that the system meets specific requirements—requirements that in the case of AXE grew out of our early experience with stored program control, or real-time computer control. As you know, they centred on handling—handling the system in design and development, in planning, production, installation, testing, and day-to-day operation. Inevitably, they turned on software, since it is software that to a very large extent provides the functional capability of the system.

For AXE, we had come up with a modular architecture, in which the modularity provides the flexibility necessary to make the system manageable. This modularity characterised both the hardware and the all-important software.

So when looking at competitors' literature and exhibition stands, we try to see the system architecture and software behind the hardware facade. Any major supplier naturally wants to show his product at the earliest possible stage, even if the design is far from complete—and it's relatively easy to exhibit some typical hardware, accompanied by descriptive literature. The difficulty for the customer is to see what, if anything, lies behind it all.

There are presentations. There are exhibitions. And there are demonstrations. Of them all, demonstrations are by far the most effective.

In 1976, we had little enough to demonstrate, but one of the best demonstrations ever was developed by Kjell Sandberg. Before he joined the select band of AXE presenters Kjell had spent several years as a trainer of marketing staff, and he was full of ideas. In the labs, we had the digital group switch working, with provisional software, and we demonstrated it whenever we got a chance. Visitors always wanted to talk through the switch, and make sure that noise did not disturb or impede speech. We elaborated the demonstration by showing what happened when the coding was less than perfect.

The switch is designed in such a way that the eight bits of each coded sample, or word, are transferred through the switch on eight parallel wires, and we use eight separate boards in the space matrix. In the demonstration, we disconnected one board at a time to show how the quality of the speech deteriorated and how distortion and noise increased. Kjell made a home tape recording which he put through the switch, pulling out boards as the recording proceeded. Extracting up to three boards (starting with the least significant) has no audible effect. Thereafter, the deterioration increases.

It may not be totally scientific, but as a demonstration, it's very successful. I've done it myself with music. It may be nonsense to say that three eighths of Beethoven is unnecessary, but Beethoven's violin concerto sounds fine with only five bits in PCM coding. May Beethoven and Nathan Milstein forgive me!

At a somewhat later date we revised another demonstration, of a different kind. At the heart of the AXE system we have the central control, the two central processors with their respective programme and data memories, working in parallel synchronism. An obvious question in the mind of a conscientious customer is: what happens to my exchange if a fault occurs in the central control? We decided that we needed to be able to demonstrate. Hans Flinck first tried out this in the Södertälje exchange. He figured the simplest way to provide an answer would be to show what happened if a printed circuit board were extracted during operation. He tried it once, pulled out a central program store board (during operation of course)—and was close to being shot by the people in the control room.

The next opportunity came up a little later, in Turku. This time we approached the subject more tactically and more tactfully, during a lunch with Veikko Tahti. He was game, and during that afternoon we tried it on the then new digital tandem.

I have since had many opportunities to run this demonstration in Turku, to many different groups of visitors. When explaining what we are going to do, it is obvious that two things are expected: a) that the maintenance staff will become hysterical over somebody interfering with the system, and b) that the system will become hysterical over being interfered with. I have seen the expectant nasty gleam in the eyes of spectators hoping for the worst.

Actually it's not very dramatic. Nothing much happens. No hysterics, on either count. If we happened to pull a board in the processor that was executive at the time, the supervisory equipment would immediately switch in the other of the pair to be the executive, while the faulty brother went into a fault-finding routine. Eventually, after some ten minutes, there would come an alarm to alert the staff and a print-out would be presented identifying the faulty (in this case missing) board.

Veikko has a well filled visitors' book today. He and his excellent staff have been most helpful and given much of their time in providing opportunities to bring visitors to Turku. The Turku Telephone Company, part of the municipal services of the city, is a small organisation by international standards. But its small size makes it easy to overview, it's easy to get an understanding of the network and the organisation to plan, build and run it. No doubt AXE has helped put Turku on the map, at least for the international telecommunications community.

Spinning the web

Behind all the centenary junketings and publicity gimmicks, we had a serious, complex and important message to transmit. If we failed to get it across, we should fail with AXE—or at least delay its progress very seriously.

But getting it across was very difficult.

Most mature disciplines develop a tradition of specialists, and telephony is no exception. We've always had the line people, the transmission people, the switching people, and a few extra crafts thrown in—radio people, power people, and others.

This applies in both manufacturing and operating companies, and on both sides the specialists form their own departments and make up highly partisan fraternities.

Ericsson has always supplied very nearly the full range of equipment that makes up the complete telephone network, and you can see the resulting compartmentalisation very clearly in the organisation charts in Chapter 1. In the parent company, for example, you can see how network material and construction, transmission, and switching were set up in separate, well-defined divisions.

The same thing happens in the operating companies. In the maintenance area, you would find separate units responsible for outside plant (equipment installed outside the exchanges), transmission systems, and the exchanges themselves. You would also find—and this is still often true—corresponding departments responsible for planning and installation: transmission engineering, exchange engineering and outside plant engineering. These departments would also have the job of defining the requirements when the company proposed to acquire new equipment from a supplier. They would do their own evaluation and purchasing, and of course each department would prepare and be responsible for its own part of a construction budget.

Over the years, several new switching systems have been introduced to the market. Each has offered, or claimed to offer, some advantage over its predecessors, and from time to time operating companies have gone through the process of selecting a new system for their networks. Between 1950 and the mid '70s, nearly every system was electromechanical. The competition was largely between the classical Strowger system, or other switches and selectors with moving contacts, on the one hand, and the newer, relay-derived systems, such as crossbar, on the other.

Crossbar made significant inroads because its maintenance costs could be proved to be lower. It's true that the cost of the equipment in the first place was always a significant factor, but with hindsight one can see that the two switching systems were more or less equivalent in cost. Perhaps crossbar cost a little more to install, but its lower maintenance cost compensated for this higher initial investment.

The point is that the systems were being compared with each other, and not with the ideal. The boundaries of the comparison process were defined by tradition. The telephone exchange was the telephone exchange. Generally speaking, any system performed the same function as any other, and what happened outside the exchange was somebody else's business. Similarly, the transmission experts evaluated and chose their own equipment. The cable people did the same. And so on. Digital switching has altered the whole process fundamentally.

The first new factor is IST, the integration of switching and transmission. IST in turn produces IDN, the integrated digital network.

And it is *integration* which has had such a profound effect on the costing of networks. The point is not that digital switching in itself is more economical than analogue switching. It is the combination of digital switching with digital transmission which has lowered network costs, and made it necessary to tackle the whole question of evaluation and analysis of costs in a completely different way. Traditionally, a combination of the most economical products and components would produce the most economical networks. Now, it could well produce the opposite.

To market digital systems, we had to move out from the exchanges in isolation and consider the total network. Digital technology influences not only the cost of a telephone system, but the whole approach to network layout and configuration. It offers a number of new solutions to classic problems.

The implications, as we prepared to approach our markets with these new concepts, were immense. We had to teach our own people a new way of life, in which telephone switching meant a lot more than telephone switching. And what was more, we had to teach our customers the same thing. We went about it by building comparison models.

The technology is covered in the second lesson from the gospel. Here, we're interested in the *financial* outcomes of integrating switching and transmission.

Figure 35 is the starting point. It shows the two switching components of an *analogue* exchange in its network environment.

Theoretically, the easiest way to bring down the cost of

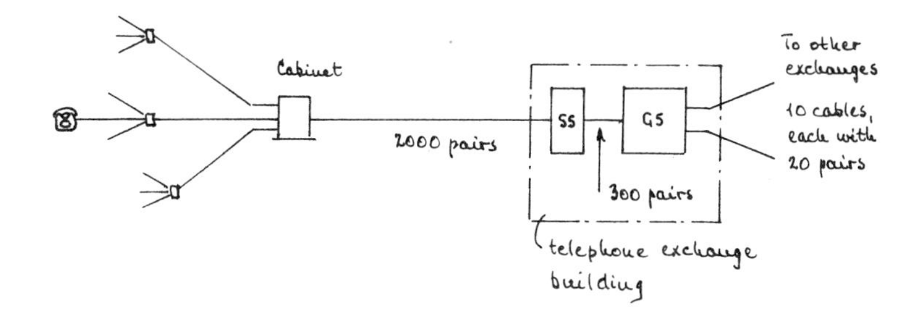

Fig. 35. Analogue local exchange with local subscriber switch.

the cable network is to move the subscriber switch out of the main exchange altogether—to bring it nearer to the subscriber. The obvious place to locate it is in the cabinet, which reduces the number of pairs in the primary cable from 2000 to 300. Network designers have been trying to do this for years—even in crossbar systems—but it's never been entirely successful. One of the problems is that the remote unit, or concentrator, needs its own housing and power supply, and these cost more than the savings made on cable.

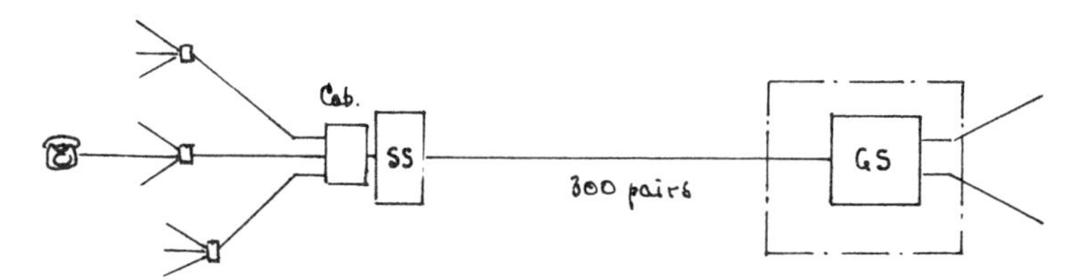

Fig. 36. Analogue local exchange with remote subscriber switch.

But suppose we now use PCM transmission for the 300 channels between the concentrator and the exchange. It will take ten PCM systems, each, of course, using only two pairs of wires. 20 pairs in all—a significant saving even on 300 pairs, and certainly very different from the original 2000 pairs. But we shall need PCM terminals. (At this point, each line has or shares a unit which converts the analogue speech signals on the line to digital code—the codecs.) The cost of these terminals is only offset if the distance between the exchange and the concentrator (and so the cost of the original cable) is pretty considerable. And we still have the cost of housing a fairly large piece of switching equipment to take into account.

Figure 37 shows the outcome. (Here, the trunks are provided with PCM as well.)

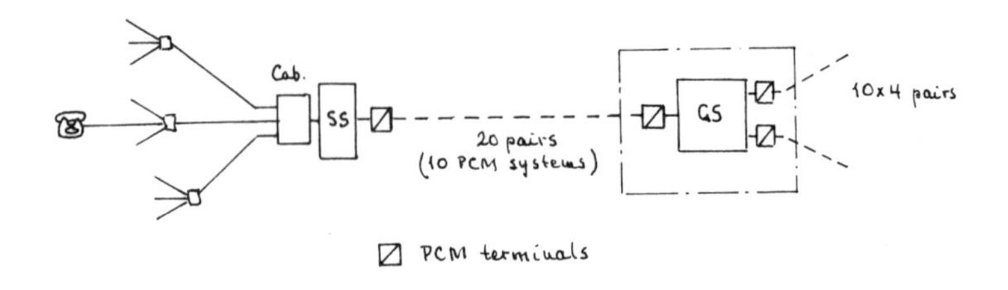

Fig. 37. The network of figure 35 with digital *transmission.*

But now, let's build in digital switching. Instantly, we eliminate PCM terminals everywhere, except where the subscriber lines enter the subscriber switch.

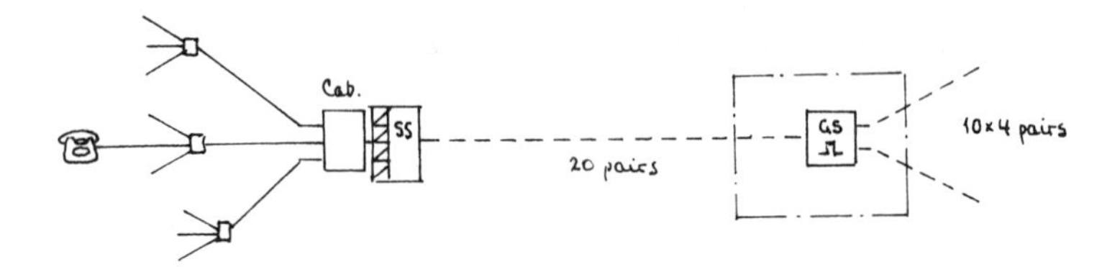

Fig. 38. The network of figure 35 with digital transmission *and switching.*

Digital switching offers an additional advantage. Digital switches are fully electronic, and take up less space—cutting building costs for both the main exchange and the concentrator.

We can now look at the *total* effect digital technology has on a network.

In Fig. 39, the cost of a digital exchange is assumed to equal that of an analogue exchange, which was true at the time.

The cost of the concentrator switch is included in the cost of the main exchange. The concentrator is, after all, simply a part of the main exchange which has been relocated.

The figures assume that digital technology has been applied over a complete network, covering an entire city, which means that fewer main exchanges are needed. (This is perfectly legitimate, since that would indeed be the case, and it naturally reduces the cost of the digital model.)

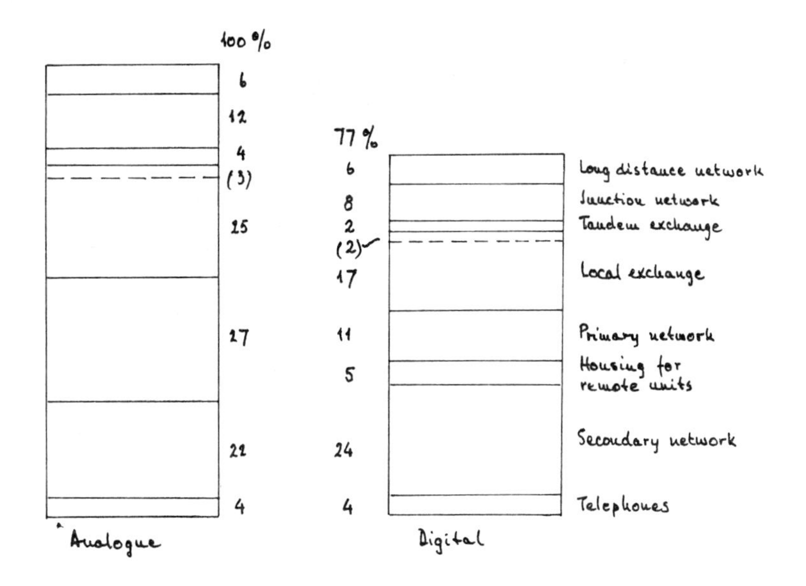

Fig. 39. Comparison of cost distribution in analogue and digital networks.

We can identify separately the savings on the primary network and the savings on the trunk network, and although we have reduced the exchange costs, we have added a cost to account for housing and power for the remote units.

The total savings are dramatic. A reduction of 20% to 25% on investment in network building costs saves 400 dollars to 500 dollars per installed line—which is a lot of money. On a fixed budget, it could mean 20% to 25% more lines installed in a year. It might even mean lower rates for subscribers!

This was the difference digital switching could make, and this was the message we tried to convey to our customers in 1976. In the early stages we could only indicate trends, only identify the direction which telephone switching was taking. We made it quite clear that it could still be some years before the full range of digital switching elements would be available. But from the beginning, we could identify many situations where a digital group switch was the elegant solution to a lot of problems; and in fact, most of our contracts have included digital group switches.

Gradually, the industry has accepted the message. And as a result, all our competitors are busy developing digital systems to replace their earlier products.

Digital AXE has given Ericsson a leading edge, and with the work going on internationally to develop the integrated services digital network, ISDN, in which every digital AXE exchange ever installed will be a component, our customers' acceptance of AXE will turn out to have been wise.

But in 1976, few people fully understood digital switching or realised its implications. Our job was as much to educate as to sell. Our competitors made the whole thing far more difficult. Relying absolutely on their older products, they became increasingly desperate in their insistence that digital switching was for the birds, or at least for the future—that the world was just not yet mature enough.

It was a mistake. You can call your competitors immature, but you mustn't imply that your customers are equally immature. Many Administrations refused to accept such an implication. They were quite ready to listen to us, and also, eventually, to order digital telephone exchanges. Most of them AXE.

The key word, however, is "eventually". By 1977, digital switching technology was beginning to be accepted as a good thing, but just *why* it was so good was by no means generally understood.

Engineers, of course, found the possibilities fascinating, and were quick to recommend digital technology to their managements as the answer to all an operating company's problems. The newspapers are full of the fears of the man in the street that engineers and technologists will take over the world, forcing unnecessary, unwanted machines onto us in their horrid and devious way. In telephony, it's exceedingly unlikely! However much the engineers like to play around with new machines, they are very much governed by the commercial rules of the game—operating-company managements are made up of hard-nosed individuals with a firm grip on the purse strings. I don't think they're particularly conservative or against innovation, but they certainly like to know what anything new is good for, and what's in it for them.

With the important message we had about the economic advantages of the digital network, we had to get through to operating company managements—which meant teaching them and their advisers the basics of the new technology. We went about this in many different ways. One of them was a book.

Within the product-planning department, one group had been working for many years writing computer programmes for planning and optimising telephone networks. They used computers to site a new exchange or route traffic through a network in the most cost-effective way, or to dimension the number of lines for each route. These programmes, based on years of theoretical work by Drs Conny Palm, Christian Jacobaeus and Yngve Rapp, amongst other doctors, had become a very valuable tool in giving the market a good service. We had now begun to use them with the new digital parameters with most interesting results. The message as we have seen, was clear—a digital network costs less to build than an analogue network. At that particular time, when most of our competitors were arguing for their analogue systems, this digital message was the best argument we had—if we could get it across.

So Ericsson produced a book, *Digital Telephony.* It was strictly non-commercial. It gave short, precise explanations of digital switching and digital transmission, how they work, and what the economic implications are. It avoided any mention of Ericsson products. The book was probably the first in its field, and quickly became a standard work in the telecommunications community. As events were to show, it became a major medium for carrying our message. And it naturally generated a lot of questions to us which led on to deeper discussions.

Spotlight on success

In February 1972, the decision that Ellemtel should press ahead with the AX proposal was taken.

In March 1977, the first pilot exchange came into service at Södertälje, in Sweden. It was analogue.

In May 1978, the world's first digital AXE exchange was cut over at Turku, in Finland.

In January 1978—six years after the original decision—digital AXE exchanges were at the heart of the largest telecommunications contract the world has ever known. This contract, for Saudi Arabia, gave AXE an unchallengeable lead as the world's foremost digital SPC system.

But in the early days, we did try the canine concept. One of the few places where things went almost to plan was Turku, in Finland.

Finland is a country with several independent telephone Administrations, and Turku, a municipal Administration, is one of them. Turku is Finland's third largest city. It is in the southwest corner, an industrial city with a long and interesting history centering on its magnificent medieval castle.

The Administration is headed by Veikko Tahti, an old friend of mine, whom I met when I was working in Helsinki in the winter of 1952/53. Like me, Veikko had recently graduated. By day, he was testing Siemens exchanges, while at the same time he was doing post-graduate work on crossbar design under Dr Svante Karlsson, chief technicial officer of the Helsinki Telephone Company. Veikko and I spent many hours discussing his crossbar design. Then I went to Mexico, and some years later Veikko became Managing Director of the Turku Telephone Company.

In the '60s, Veikko placed an order with Ericsson for an AKE exchange. It went in smoothly, and performed well, and in 1969 Veikko placed a follow-up order for an AKE 12 combined local/tandem, to act as the basis for further SPC network extension. AKE 12 was still in the early stages of development, and it would be quite some time before it could be installed, but that fitted reasonably well with Veikko's planning timetable.

Then, in 1972, we decided to put all our resources behind the AXE programme, and to discontinue the development of AKE. Mr Tahti is a man of great charm and understanding, but at the same time he is a tough customer and a hard negotiator. Not surprisingly, he was one of the first customers we approached, early in 1973, and we tried hard to persuade him that he would want AXE, rather than AKE, in his system. Even though that meant a slightly later delivery date.

Veikko eventually agreed, and placed the first formal order for AXE we ever received, a local exchange for 4000 subscribers. Turku now has an honoured place in the history of AXE. It has the first AXE exchange ever installed outside Sweden, and the first digital AXE exchange ever installed anywhere in the world. If only Veikko hadn't spent the years between 1973 and 1978 pointing out to visitors the large space waiting, waiting, waiting for his Ericsson equipment.

But Turku was an exception. Australia was more typical, a dour struggle between Ericsson and ITT.

The struggle for Australia goes back a long way. Tough, knowledgeable, experienced, professional, the Australian Post Office (now Telecom Australia) is universally respected. In the late '50s, Australia standardised on Ericsson crossbar, which was manufactured by our own company, EPA, in Melbourne, and under licence in Sydney by the British-owned Plessey, and also by STC, the ITT subsidiary. For years, Australia seemed tied to Ericsson, a natural for the canine concept. And in Stockholm I was in charge of technical and marketing liaison with this (as we thought) happy, trouble-free market.

In 1969, the big, bad wolf huffed and puffed and blew our house down. The year before, the Australian Post Office had issued an international invitation to tender for a large transit exchange, Pitt, in Sydney. It was a sign of the way things were to go in the '70s. In their specification, the APO identified a number of requirements generated by the surge of growth in subscriber trunk dialling; requirements for which SPC was the natural choice. With EPA, we spent several months putting together an offer of the AKE SPC system. We handed it in in January 1969.

The only other serious contender was ITT, with the Metaconta 10C system. In September, the APO announced that it had chosen Metaconta.

The effect of this blow on the decision to develop AX has been described earlier. From the marketing point of view, what happened next?

Our immediate response, in December 1969, was to send a delegation to Australia to present our progress with the AR system.

Nothing happened. We continued to supply crossbar. No SPC.

Two years later, we repeated the exercise, in December 1971. This time, Björn Svedberg headed the team as the man then responsible for the ARE development project.

Nothing happened, except that a highly tentative schedule for the possible introduction of ARE emerged.

In 1972, however, the Australian Post Office made an important decision. For a large local exchange, City North, in Sydney, the APO chose the Ericsson ARF crossbar system without SPC. We deduced that at least Metaconta would not be the automatic choice for future local exchange applications.

We were right. In January 1973, the Administration issued a specification for a system to add SPC functions to Ericsson crossbar, and in March it invited us to prepare proposals for two local exchanges (Salisbury and Elsternwick). With EPA, we responded in June with an offer of ARE 11 pilot installations.

In the autumn of 1973, the Pitt Metaconta exchange which had been ordered in 1969 was due to come into service. It didn't, while in January 1974, to our great satisfaction we were awarded the contract for Salisbury and Elsternwick.

We felt that at least our honour was avenged, but by now, over two years after the commitment to AX, we had something more important than our honour to think of.

We knew already that the Australian Post Office was planning to make a long-term system choice for local exchanges, and that a specification for international tender would go out in 1975. Our hold with crossbar had been broken. We had something to offer with ARE, but we knew that even ARE couldn't meet the APO's long-term needs.

We also knew, it's true, that the APO was insisting that any system offered should be proven; that exchanges of the type proposed should already be in service. We could not, of course, offer anything of the sort with AXE. But we were convinced we had a system worth considering, and our people in Australia had persuaded the APO to listen to a presentation of AXE and to discuss the coming tender.

In November 1974, we sent a delegation to Australia to talk, at last, about AXE.

We gave it everything we'd got. The team was headed by our executive vice president, Fred Sundquist. With him were Björn Svedberg, Kjell Sorme, Ingvar Rhedig, and I. We met several groups of people from the APO. We answered innumerable questions. Svedberg and I in particular developed an act—a certain style of back-chat and mutual insults which greatly appealed to the Australian sense of humour.

Most important of all, Fred finally reached agreement that we should be included in the invitation to tender. And in January 1975, when the tender invitation duly arrived, we set to work to prepare a proposal.

The rest of the story took a long time to live through, but can be short in telling.

In May of 1975, the ARE exchange at Salisbury was successfully cut over. In July, we submitted our AXE proposal. 1975 went by.

In January 1976, encouraged by Salisbury, the Australian Post Office decided to install ARE 11 in large quantities. In September, Elsternwick came into service, and—great progress—the APO issued a press release to announce that only AXE and Metaconta 10C were still in the running to become the future standard system.

We left nothing to chance. In November 1976, another delegation left Ericsson for Australia to discuss AXE. The climax was close. We'd been working on our own pilot exchange at Södertälje for nearly a year, and commercial cutover was scheduled for the Spring of 1977. The Australians wanted a proven system. We wanted the Australians there for the cut-over. They arrived in February 1977, and inspected the Södertälje exchange the day after cut-over, on March 2nd. It worked.

On September 13th, 1977, the Australian Administration, backed by the Government, announced that Australia would standardise on AXE for local exchanges. After eight years the dogfight that began in 1969 was over.

But life was still deadly serious.

Between the cut-over of Södertälje in March 1977, and September 1977, when the Australian Post Office announced AXE as its system choice, there were other proposals to be made.

In June, I was in Buenos Aires making presentations to the Argentine Administration, ENTEL. There were three of us from Stockholm, plus a fourth from our company in Spain, and we had a full programme.

During the final lunch with ENTEL, a call came through for me from Stockholm. It was Hans Flinck, asking if I would take charge of the preparation of part of a proposal for Saudi Arabia. A specification had been issued with an invitation to tender, and the proposal had to be in by September 27th. Three months.

I had a week planned with Janet in the west of England, and it looked as though it would be 1977's holiday. My consolation was my final week-end in Argentina. It was quite an experience. We were invited to go up to Mendoza, a large city below the Andes, where LM Ericsson still runs the telephone system. (In fact, we still have two operating companies in Argentina, serving 300,000 subscribers—left-overs from the old days, when we and other manufacturers were building our businesses on operating concessions.) We had a good time in Mendoza, with visits to the telephone exchanges (exciting to telephone engineers), a local winery, and restaurants with superb international entertainment. Above all, we took a trip into the Andes by car. We hit snow as we gained altitude, and naturally we got stuck in it, and naturally one of the cars broke down and had to be mended with a piece of string. And naturally, as a bird-watcher, I had set my heart on seeing a condor. I had instructed my companions to keep a sharp lookout for large birds, and the kidding was continuous; childish, the sense of humour some people have! But at last, as we lined up along the edge of a precipice for some necessary relief, we spotted a big bird. Sure enough, with my field-glasses I was able to identify a condor.

It was a great occasion for me, and my friends helped me to celebrate. We sang the well-known Andean song "El condor pasa", and we sang a Swedish translation. I shall not record the Swedish translation here. I'll just comment that even telephone engineers sometimes show a remarkable lack of culture and good taste.

From Buenos Aires to England, my week's holiday with Janet, and then Stockholm.

The Ministry of Telecommunications of the Kingdom of Saudi Arabia had issued an invitation to tender for the largest project ever in the history of our business. The specifications had been prepared by Arthur D. Little, well-known consultants, and embraced the entire job of extending and modernising the Saudi network with nearly half a million lines. The project covered SPC exchanges, telephone sets, and transmission systems, plus the installation of all equipment, including the building of the cable networks and the construction of new exchange buildings. In addition, operation and maintenance were to be provided for five years. And the first phase was to go into service by one year from the date of the order.

So much for the canine concept of small pilot exchanges running as field trials here and there throughout the world! The *only* AXE exchange in existence was our nursery exchange in Södertälje.

We had tried to get the Saudi Ministry to consider ARE, a proven, manageable, system, but without success. It had to be AXE, or nothing at all.

L M Ericsson had been active in Saudi Arabia since 1964, We had supplied and installed the existing network of some 200,000 lines of crossbar, which covered only the major cities. In the lean years which followed 1972, when we had nothing very new to offer, Philips, of Holland, had come to an agreement with the Saudis that they should supply the next, very large, extension to this network. All that remained was to agree on the price, and the signing of the contract.

The project was so large that pressure built up for a public tender competition. The "generous" contract for which Philips hoped did not materialize, and Arthur D. Little was called in to prepare the tender specification.

At Ericsson, things were humming. The Saudi Ministry had recommended that partnerships should be sought to reduce the risks for both sides of such a large undertaking. After hectic discussions, Ericsson and Philips agreed to join forces, and later a third partner, Bell Canada International, was found to look after the network operation side of the proposal.

Hakan Ledin was now in charge of X Division, and he set up a task force. It was headed by Ove Ericsson, who had two groups reporting to him. One group was concerned with installation, and all the logistics of the project. The other, my group, was to work out the technical proposal. All of us had put together plenty of proposals in our time—they're just part of the job. But this time, we had two new factors—the proposal had to be a joint production with Philips; and it had to dove-tail with a separate proposal from Bell Canada.

It was the beginning of July, when Sweden closes down for a month's holiday. We couldn't draw on Ericsson's normal resources, so we had to call on people from a number of different departments and build up a temporary project organisation. I was not the only person to miss or truncate a holiday in 1977. In a short time, I had assembled as fine a group of people to work with me as anyone could hope for.

Sten Henschen had been working with Saudi Arabia for several years. He knew the existing installations well. Kjell Nilsson also had a great deal of experience of the Arab countries. Kevin Casey had just come back from Brazil, and was waiting for a worthwhile assignment. He became a powerful force in making the technical aspects of the proposal concrete and coherent. Jan Nordin was in charge of the X/S department at the time, the department which supplied a range of technical support services to the sales departments. He and his people handled many of the studies on which the planning of the network was based. Jan also took charge of the operation and maintenance equipment element of the proposal.

There were many more. None of them had to be dragooned into the Saudi project group. We were all aware that the future of the company depended almost entirely on the success of AXE. It wouldn't be the end of the world if we didn't get the Saudi contract—but if we *did* get it, and if AXE performed as it should, L M Ericsson would be unstoppable for a decade. It was as simple as that.

It was also, of course, a fascinating and thrilling project with which to be involved.

To build and operate a national network it is usual to start with a series of fully-developed fundamental or basic technical plans. Such plans normally include the national numbering plan, the national transmission plan, the routing and switching plan, the operating and maintenance plan, the charging plan (tariffs, and the charging of different types of call) and, for a digital network, the national synchronisation plan.

The plans define, if you like, the sort of system which you are setting up, what it will be like to operate, how comfortable life will be for subscribers. There were no plans in the specification, so we decided early on to develop them. In our collaboration with Philips, we should then have a set of guidelines and a common framework, within which we could equip and dimension and engineer the network components.

Our first decision at Ericsson was to use AXE with digital group switches. We were convinced that this would give us a cost-effective proposal with considerable technical competitive edge.

So when a group of us travelled to Hilversum in Holland for our first technical discussions, we had with us the preliminary suggestions for a full set of basic technical plans, and we could talk through the implications and opportunities of developing a network which would incorporate digital switching.

Such discussions could be comparatively high-minded. But at an even higher (or lower) level, Ove Ericsson and Ericsson's senior management had to discuss with their Philips counterparts the very charged question of who should supply what. I've been told that this exchange often became quite heated before agreement was reached.

The outcome was that the supply of switching equipment would be divided strictly 50/50, with Ericsson's AXE proposed for the large local exchanges and for the tandems, and Philips's PRX for the remaining, smaller exchanges.

Ericsson would offer telephone sets of a new, fully-electronic design. Philips would be in charge of the bulk of the PCM transmission systems and cable.

Exchange buildings would be designed by a Philips subsidiary, and built by an outside contractor who would also handle cable network construction.

There were many trips to Holland during July and August, and the Philips people spent many weeks in Stockholm. It's a marvel that we managed to put the whole proposal together in the short time available, but we did. There were snags, of course, but we worked them out.

The stakes were so high for both of us that comparatively minor issues became wildly inflated. The tender specification, for example, set down in detail how the proposal should be constructed. The order and numbering of the chapters; how the different sections should be marked and indexed; and so on. This became the basis of some heated arguments, in which I was involved. At an early stage, I had formed the idea that we needed an executive summary of the total proposal, a nicely-presented document in the form of a booklet of some

30 pages, outlining the main features of the proposal in non-technical language. The Little proposal had not asked for such a document, and the person at Philips in charge of producing the total documentation could not accept the idea. He claimed that we would be disqualified if we broke the rules like that. However, I'd become friendly with a young Philips engineer, Peter Calis, who shared my belief in a summary, and together we wrote it. It was agreed that it could be inserted at the beginning of the first binder, but it must be printed in the approved standard style, without frills. Secretly, in collusion with Peter, I had it printed in Stockholm, with colour photographs, diagrams, the lot. We spent an unpleasant Sunday with the Philips project management, revising, changing, and "improving" the contents, but it was passed.

At the final gathering, we were complimented on the good-looking booklet. (There had been a last-minute flap, when somebody discovered that on a map in the booklet the name of one city had been misspelled. Of all the cities in all the world, it had to be Mecca. We spent half the night pasting tiny pieces of paper over our own version—Meccca.)

It had been agreed that the final preparation of the proposal should be done in Hilversum. A large group of us spent a fortnight in Holland. With four of our best typists and their word processors, we were to go through every page, make corrections, check against the original specification for omissions, repair them, and finally assemble the complete package.

It was quite a package—42 binders, 6,000 pages, an impressive stack 2.7 metres high. We'd been assigned an unused laboratory, and as the days went by we spent more and more hours in this building, and in the printshop. I have a photograph taken on one of the last nights—a long table down the middle, with people reading, arguing, drinking coffee. In the background, a couple of bodies are asleep on top of some tables. It was war, it was fun, it was a terrific experience— though I'm not sure that at the time many of us believed we had a chance of actually getting the contract. We delivered the proposal to the Ministry on time.

The opening of the offers was public. Our tender price was the lowest. But we all knew that the proposals were so large and complex, and presented so many different solutions and possible combinations, that the opening prices could not be the basis for decision. We should have to live through a period of evaluation.

On December 13th, 1977, the Ministry's decision was announced, and on January 25th, 1978, the contract with

Ericsson-Philips-Bell Canada was signed. It was the turning-point for AXE.

The canine concept was finally dead.

But though the Saudi project was the largest single order ever to be placed, Saudi Arabia was not the largest single market for AXE.

The market with the most AXE installed is, oddly enough, France. Yet Ericsson hasn't supplied an exchange to France for years!

It happened like this.

In 1975, just after we had submitted our AXE proposal to Australia, the French Administration issued an international tender invitation for a new stored prgoram control telephone system.

France was already making use of CIT-Alcatel's E10 system. E10 was designed for the rural areas of France, for its small towns and villages. Now the Administration wanted to introduce SPC on a wider scale, in the major cities.

Through the Ericsson company in France, Societe Telephones Ericsson, STE, we were already working with CIT-Alcatel. We had started a joint research company about a year before the tender invitation. Its aim was to develop a large-scale digital switching system, by pooling Ericsson's expertise in high-capacity control with CIT-Alcatel's experience in digital switching.

Progress had been slow.

The proposal had to be submittd in September 1975, and the situation was tricky. The French government required tendering companies to be French, and manufacture of the system chosen would have to be in France. Since STE was already collaborating with CIT-Alcatel, STE's proposal for AXE was submitted by CIT-Alcatel—who also handed in a second proposal, based on Nippon Electric's D-10 system!

The other major French company to submit a proposal was Thomson-CSF. Thomson had no switching system of its own, but had set up a license agreement with Northern Electric Company of Canada (which later became Northern Telecom). Their offer was based on Northern's SP-1 system.

It should be pointed out, perhaps, that the French PTT was specifying an analogue system.

All very confusing, you may feel. But after a period of evaluation, the French made a simple, straightforward choice. They chose Thomson-CSF as the second French supplier. True, they rejected the Thomson SP-1 proposal, but what of that? Instead, they chose AXE, but to be supplied by Thomson, and not by CIT-Alcatel!

The outcome of it all was exceedingly intricate negotiation. Thomson acquired a licence from Ericsson to produce AXE, and also bought STE. ITT was forced to sell off part of its French operation—two companies: LMT (Le Materiel Telephonique); and LCT (Laboratoire Central de Telecommunications). With LMT, Thomson acquired the French version of ITT's Metaconta system, the 11F.

The order was placed in December 1976. It was the first large contract for AXE, and it was not with Ericsson. We would manufacture the first few exchanges in Sweden, but thereafter manufacture would be transferred to Thomson in France.

To some of us, the loss of our presence in France, where we had been in business for over 50 years, seemed a high price to pay. But as a reference, France was of inestimable value to us at the time. AXE had won acceptance by a major European Administration in an international competition.

The order for the first exchange, Orleans, was signed on December 2nd, 1976. The exchange was handed over by Thomson on September 30th, 1978. It was officially accepted and taken over by the PTT in May 1980.

Such time-scales are by no means unusual, and it often pays to play a waiting game. France today has more lines of AXE in service than any other country.

(This happy situation will not continue forever. The original specification was for analogue switching, and every AXE line in France is still analogue, because the French PTT intends eventually to replace AXE with Thomson's all-digital MT 25 system.

Turku, Australia, Saudi Arabia, France. From the almost cosy to the most complex. From the comparatively predictable to the Saudi thunderbolt. From discussion in Finland in January 1974 to a 476,000-line contract in January 1978. No system of comparable novelty ever moved so far, so fast, from a standing start.

But even the contract from Saudi Arabia was not so much the end of a process, as the beginning.

Success on Record-I

Like the other manufacturers in the telecommunications business we make reference lists. They are an important detail in demonstrating to other customers how well accepted a product has become. Here again we use the number of lines as the quantitative factor, rather than price, and incidentally we are always very careful to include only those installations for which we have firm contracts. In December 1976 we made our first reference list for AXE and it looked like this.

December 1976

Country	Exchange	Multiple Capacity On Order
Finland	Turku	13 000
France	Orleans	13 000
	Nantes	10 000
Sweden	Södertälje	3 000
Total		39 000

Puny, you may say—and we thought so, too. Still, we were proud to show it during our presentations; after all, it was a new system, and we had so far been very selective in our marketing. And the two French orders were of course only the first two contracts; the French had made a system choice of AXE, and Orleans and Nantes would be followed by many more installations.

By May 1977, the reference list looked a good deal healthier. For one thing we had cut the first AXE exchange, in Södertälje, into service on the first of March that year. And the Swedish Telecommunications Administration had placed contracts with us for ten new exchanges, in various cities, plus an extension of the Södertälje installation.

There was a story behind this first large order from our compatriots. It was something of a political deal. At this time, 1976/77, we were feeling the economic squeeze of the world recession, a result of the first oil crisis. Orders for crossbar had dropped off dramatically, and we had over-capacity in our Swedish factories. And one did not need a crystal ball to understand that there was little chance of crossbar sales picking up if and when the depression passed. The telecom world was changing. In future our markets would be wanting computer controlled systems and, as we ourselves were arguing, these systems would to an increasing degree be digital. SPC and digital telephone exchanges require much less manpower in the manufacturing process. Even if the market picked up again we would have too large a production facility. So the company announced that it would have to close one factory, in the south of Sweden. Such closures are not undertaken lightly; they must be negotiated with the unions, and in this case the government was called in—with the eventual result that the administration was directed to place an order for 240,000 lines of AXE with L M Ericsson, and L M Ericsson agreed to keep the factory going for another three years.

May 1977

Country	Exchange	Multi Cap. In Service	Multi Cap. On Order
Finland	Turku		13 000
France	Orleans		13 000
	Nantes		10 000
Sweden	Aspudden		34 000
	Boras		40 000
	Gavle		36 000
	Jonkoping		10 000
	Karlstad		10 000
	Masthugget		38 000
	Savedalen		10 000
	Södertälje	3 000	32 000
	Tumba		10 000
	Vasteras		10 000
	Ostermalm		10 000
Total		3 000	276 000
		279 000	

By December 1977 the reference list had become quite long and we resorted to two overhead slides when showing it. We also introduced a separate column showing tandem and transit exchanges. Incidentally the heading "multiple capacity" is a sort of leftover from the electromechanical era. The numbers listed denote quite simply the exchange size, or capacity, in number of subscriber lines and trunk lines respectively. The list from December 1977 follows.

December 1977

Country	Local Exchanges	Multiple Capacity In Service	Multiple Capacity On Order
Finland	Turku	. 4 000	9 000
France	Argentueil		12 000
	Chalon s/Marne		7 000
	Chartres		12 000
	Joué les Tours		9 000
	Longwy		8 000
	Metz		10 000
	Nancy		12 000
	Nantes		10 000
	Nevers		8 000
	Orleans		13 000
	Strasbourg		6 000
	Versailles		8 000
Kuwait	Ardiyah		10 000
	Mina Abdullah		10 000
	Mishrif		10 000
Spain	Atocha		10 000
Sweden	Aspudden		34 000
	Boras		40 000
	Gavle		36 000
	Jonkoping		10 000
	Karlstad		10 000
	Masthugget		38 000
	Savedalen		10 000
	Södertälje	3 000	32 000
	Tumba		10 000
	Vasteras		10 000
	Ostermalm		10 000
Yugoslavia	Vrapce		10 000
Total		7 000	414 000
		421 000	

December 1977

Country	Transit Exchanges	Multi. Cap. In Service	Multi. Cap. On Order
Denmark	Alborg		9 000
Finland	Turku		480
Total			9 480

Quite a lot had happened during the year. For one thing, the French orders had been coming in as expected, and the first phase of the Turku exchange had been put into service. And we had gained some new markets—Kuwait, Spain and Yugoslavia, with orders for local exchanges. I heard about the Kuwait order while in Teheran. I should say I read about it, because we couldn't get through on the telephone and had to have a telex conversation. I was in Teheran with Ingemar Nilsson, who was in charge of sales for that market, and the man at the other end, maltreating the telex machine, was my good friend Nazif Khalidy, our regional sales manager. Ingemar was shaking with nerves before we got the garbled message clear. When we arrived in Kuwait a couple of days later we had a marvelous feast in Nazif's house, supervised by his charming wife Aida.

The Atocha order for Spain was a sort of trial installation for the Administration to become acquainted with the new system, and the order for Yugoslavia was the first part in a licence agreement we had signed with the Nicola Tesla Company of Zagreb.

We had also got two orders for transit exchanges, in Denmark and in Finland. The significant point about these first two transits (Turku is really a tandem exchange, but from a functional point of view that is the same thing as a transit) is that they were our first orders for fully digital AXE exchanges, and this indeed was a breakthrough. Especially so since the local exchanges for Kuwait, Spain and Yugoslavia were also specified with digital group switches (and later the Swedes would convert theirs, too).

So 1977 ended in a major key. Our strategy of introducing digital switching at an early stage was bearing fruit. AXE was becoming known and respected. And not even shown on the list were the orders for some 160,000 lines of AXE for Saudi

Arabia, announced during that same month of December 1977.

AXE was making waves in the telecommunications world, and naturally the competitors were now aware of what was happening and were beginning to look around for means of meeting the challenge. We realised this, of course, but whatever they came up with we were fairly confident that we had a technological lead over most of our traditional competitors—a lead we estimated to be of the order of about two years.

MARKETING AND THE ETERNAL TRUTHS

This section is a joint product. Jeans is a modest dogooder, by no means certain that anybody is interested in AXE, engineers, or John Meurling. He suggested that people ought to be rewarded for reading the book. The only reward he could think of was a check-list. His mind works that way. The only check-list he could think of was one which would guarantee success in the marketing of technological innovation.

Meurling, on the other hand, knows what he's talking about (which Jeans often doesn't). So Jeans has produced a theoretical check-list, which Meurling has validated against reality.

In fact, this section grew out of the question we asked ourselves with monotonous regularity—why did AXE succeed?

Advertising agencies are in a better position to do comparative analysis of marketing than most organisations. Unlike consultants, they are long-term participants and not merely voyeurs. Over the years, dozens of case-histories present themselves as clients. And the work of any worthwhile agency is 90 percent observation and analysis, 8 percent producing solutions to problems, and 2 percent selling those solutions to clients. It's in the nature of a good case-history that its analysis should provide truth, of a kind. That kind is usually a model, a descriptive framework with time as an important component. Time can be regarded simply as a record of change.

The point of a model is a prediction. Any new situation can be compared with the model. If it is very like the model at a given point in time, then (the theory runs) the odds are that if you replicate over real time the changes which take place over time in the model, the results which happen in the model will also happen in real life.

And so, in fact, they probably will, if the model is a good one, and if the live situation really is like it.

By any standards, AXE has been a marketing success. Is it a good case-history? Can a model be derived from it, which will be relevant in other situations? If so, what are the elements of the model which are significant? And can the model be validated by comparison with other real-life case-histories?

It's worth a try.

Any theorist can turn up half a dozen definitions of marketing in as many minutes. Most of them have two things in common—that the process starts with a definition of a market's wants or needs; and that the objective is profit.

Both elements are certainly present in the AXE case-history. In the section on the requirement specification, we can note the decision that the requirement specification should be "written by the markets". And AXE was certainly profitable after the initial period.

But many, many people set off on what seem to us two entirely wrong directions from those two starting points. To discover what the market needs or wants, they commission market research. And to control the profit, they install an accountant at the head of the project (or the company) and exercise control through very rigorous management accounting procedures.

Sound market research and tough financial control (not so much to produce profit as to avoid out-running resources) are, in our view often necessary but never sufficient conditions for the successful marketing of technological innovation on a large scale and over a long period.

The most significant elements of any model based on the history of AXE would, we believe, be something like these.

— A leader or a team—preferably a team—with a successful, recent, personal working background in the engineering and technology of the product.
— A trial-and-error history of failure to achieve what the market wants; a "learning curve" traversed by the supplier *and* the market.
— The depth of technological resource to provide an almost inexhaustible supply of alternatives to failed attempts.
— A fairly high level of continuity and security of employment, perceived authority, and honest niceness diffused among staff.
— A feel for timing, and for luck.

We believe those conditions are to a greater or lesser degree necessary. We're not sure they're sufficient.

It's perhaps worth considering the elements of the checklist individually.

The importance of leadership with an engineering background

At some point, most technology-based companies have a financially frightening time. (Ericsson is no exception. The period following Kreuger's suicide in 1932 was financially terrifying.) Some companies have several of these times.

Very few companies escape them altogether. Often, the cause is success. The company finds itself over-trading, or even investing in projects with good long-term prospects and no short-term finance. Only too often, the result is predictable—the accountant and the miracle management information systems. These are not bad things in themselves, but they *are* bad if they become the soul of the company. Yet so often they do, while the company's real raison d'être becomes the subject of an R&D department (to which the founder may well attach himself).

The outcome is inevitably a reduction in technological and engineering impetus. The emphasis switches to margins and return on dollar invested. The Boston model, or something similar, is introduced. The company becomes a mechanism for making money rather than products. It shows fine financial results—so fine, that it attracts take-over or nationalisation bids, and everyone wins. Except the technological innovators and the public, which might have benefitted from their innovations.

These also often become the companies which rely on market research to produce the equivalent of a requirement specification. In a company where the people who represent its dynamics have operated intimately with customers and markets, requirement specifications almost write themselves. Market research may validate them or quantify the demand for the product, but the functional requirement is almost obvious. (Almost, not absolutely, and frequently there is conflict between apparently mutually exclusive requirements.)

By great good fortune, Ericsson avoided the normal outcomes of a financial fright after 1932. Its financial recovery was managed by Marcus Wallenberg, the legendary Swedish banker. Exceptionally, Wallenberg understood that technology must be the driving force of a technological company. From its beginning to the present day, the senior posts at Ericsson have been filled by engineers.

And maybe this fact helps to explain why AXE was the first full-scale digital system on the market, and why it still has a substantial lead.

The man who never made a mistake never made anything

We've outlined the story of the attempts to bolt SPC onto crossbar (ARE, AKE).

The systems worked—and still work—well enough to be worth installing. They, and their analogues from our competitors, represented the best solutions of the world's finest telecommunications engineers to the problems as they perceived them. The Administrations who bought them were making sensible decisions. Without the experience they provided, the advance to AXE could not have taken place. No amount of laboratory simulation could have demonstrated to Ericsson *or the Administrations* the handling-costs penalty imposed by the first and second generations of SPC in real operation.

The best is the enemy of the good—and if it suppresses the good from which it might eventually be born, it is self-destructive.

Any area of technological innovation in which the boundaries between purchaser and supplier, engineer and salesman, are blurred is probably on the right lines and likely to be fruitful. The trick is to prevent hardening of the arteries as the business moves from youth to maturity.

Depth of technological resource

Trial-and-error learning is a protracted, expensive process. It can also be disheartening; it takes perspective to see that the process is necessary. At the time, any setback in the field seems like the death of the underlying concept.

If it is a one-man concept, that man needs both resilience and conceptual fertility to overcome the setback. A team is a safer bet. Somebody will see the setback as an indication of a flaw in the concept, and an opportunity for salvage and improvement by a change in technological direction, which he just happens to have in his bottom drawer.

A team provides stamina. It also improves the chances of a genuine creative advance. The "Eureka" approach is glamorous and admirable, but unpredictable. Bankable creation is usually a process, not a sudden brilliant illumination. It's a series of "Yes, but what if . . ." "Suppose we try . . ." "That won't work, but this might . . ." conversations. The impetus for the creative leap may come from the market (Ericsson in the drive for AXE, the techniques from the engineers (Ellemtel).

Continuity, authority, honesty

We've said that continuity of employment is a Swedish tradition. We believe it has two happy outcomes.

The first is that it allows mistakes to happen, without crushing penalties. The stressful environment, with the spectre of the ultimate sanction stalking the sleepless nights, is supposed to be a powerful motivator. It's our guarded belief that good people create for themselves the amount of stress they need, and that the artificial injection of stress does not transform the sluggard into the ant. In such an environment the error element of the trial-and-error process can be seen as part of progress.

This does not, of course, mean that innovation can be uncontrolled. It does, perhaps, mean that management may be largely enabling—consisting of the clear description of goals, with frequent reinforcement, and the provision of enough information to allow an engineer to generate his own relevant stress.

The second happy outcome of continuity of employment is continuity of contact with the market. As a senior man within Ericsson, John is in contact with senior men in telephone administrations who were friends when he, and they, were little more than boys. This is plainly a congenial atmosphere within which to generate sales. It's also very valuable as an environment in which to accumulate continuous market intelligence. Ericsson engineers spend a great deal of their careers, often years at a time, abroad. The relationships they build up, and the depth and intimacy of their involvement with markets, certainly enabled the requirement specification to be "written by the markets."

But agreeable fools do not make valuable representatives. Successful Ericsson men abroad must offer authority. The markets must find them reliable and trustworthy. The authority of any executive is a function of three facets of him—his personal life; his professional performance; and his proprietorial attitude to his company ("I must act as though I owned Ericsson personally"). Ericsson engineers are neither Einsteins nor Schweitzers, but overt training, exposure to Ericsson tradition, and nationality, have combined to give them a reputation for reliability. In general, markets expect Ericsson engineers to provide soundly-based help. In general, they are not disappointed.

Timing and luck

Many things which look like luck, aren't. Luck is the genuinely unpredicted, unpredictable, element which improves the prospects of a project. There's nothing more to say about it, except that it's present in an astonishing number of case-histories of success. In the story of AXE, it was the oil crisis, which bought time and allowed Administrations to catch up.

And time, and a feeling for time, are always significant elements. Ask creative people, in almost any environment, what their greatest enemy is, and they will usually name time. From 1970 to 1975 the shadow over AXE was time—the knowledge that for several years Ericsson would have nothing new to offer; the need to maintain impetus without disheartening engineers or forcing them to scamp work; the difficulty of estimating how long AXE would take, and should take. The anxiety is vividly shown in the stream of forecasts and time models produced between 1972 and 1976.

We would rate the ability to handle time—to be patient when patience is appropriate, to hustle when hustling is needed—as a significant element in the model.

In the story of AXE, these elements came into balance. They were necessary; were they sufficient to produce success? We're still not absolutely sure. We do know that market research and rigid financial control made little contribution.

We're still examining some other elements—quality, for example—and the history of AXE is far from complete. However, it's too late for disaster to strike AXE, and we think the elements we have outlined can claim credit for that much at least.

LAS MARIMBAS

The marimba is the popular traditional musical instrument in an area stretching from Southern Mexico down through Central America. It is somewhat like a large xylophone with a range of different-sized wooden blocks, underneath which are mounted individual resonance chambers made out of gourds. Several musicians line up along the marimba and produce the characteristic music by hitting the blocks with light mallets.

In the autumn of 1980, the Guatemalan Telephone Company, GUATEL, had invited three manufacturers, NEC, ITT and Ericsson, to a three-day seminar. We were asked to present our system products, one day for each company. NEC was to have the first day, ITT the second, and we would do our stuff on the third day, a Friday. We had planned a day's programme focussing on the digital network, presented by Walter Widl and myself.

I arrived on Wednesday night, a day late because of some booking mistake, to find my Ericsson friends somewhat agitated. The ITT contingent, six engineers from SESA in Spain, had made a request for more time, and so we had been asked if we would agree to reducing our programme to just half a day. Naturally, my colleagues were not inclined to go along with such a preposterous suggestion—but after talking it over, we decided to accommodate our ITT friends, concentrate our gospel into about two and a half hours, and gracefully leave the floor to the competition for the Friday afternoon.

The seminar had been arranged in an old building in Antigua, a charming colonial town about an hour's drive from Guatemala. Antigua was once the capital of Guatemala—under the Spanish—but was abandoned after severe earthquake damage. In the clear sunny weather, several volcanoes around the town provided an impressive backdrop.

GUATEL's own marimba band, six men plus a drummer, was on hand to provide music during intermissions, and before and after the day's programme. We all enjoyed the breaks with marimba music, small cakes, and strong Guatemalan coffee.

ITT's day started with an introduction of Network 2000, a concept built around their System 12 digital switch, and then proceeded in high-speed Castilian Spanish to cover all aspects of all their products. A very full day. ITT gave us all a good lunch in the patio of a local restaurant.

For our programme on Friday morning we had prepared a concentrated presentation. Walter and I took it in turns to discuss essential questions of economy, planning, and operation and maintenance. We were an efficient team by then and, I believe, had found the right mixture of technical, economic and maintenance information, with a suitable infusion of light humour. I ended the presentation with a reference to yesterday's very interesting Network 2000 concept, mentioned how thrilled we had all been, and pointed out the difference—what Ericsson had presented might be termed Network 1980! I then invited everybody to lunch—

including the marimba band.

Lunch had been arranged in the garden of the main hotel. Walter and I were pleasantly relaxed with our job done, and our aim was for this same relaxed mood to permeate the whole group. In this I believe we succeeded. There was local wine and beer and a typical Guatemalan buffet lunch. A general feeling of goodwill was enjoyed by all. Over coffee and wine, accompanied by the marimba, we sang songs of great deeds and great loves.

At 3 o'clock, about half the congregation ambled back to the seminar building for the final ITT session, and by 5 o'clock the rest of us had made it. We all wanted to be present for the final ceremony and the last marimba solo.

CHAPTER 4: BACKSWING AND FOLLOW-THROUGH

INTO THE WORLD MARKETS

Swedish Pastry

The announcement that Ericsson-Philips-Bell Canada had been chosen as the contractors for the expansion of the Saudi Arabia telephone system came on December 13th, 1977. Janet and I had arrived in Mexico City the day before and I got an early morning telephone call from Stockholm. It was Hans Flinck with the news. Hans is a cool type, not given to excessive outbursts of emotion, and our conversation was short, matter-of-fact and to the point.

We had made it. We had succeeded in landing the largest telecommunications project ever awarded. We had succeeded in beating two powerful giants, Western Electric and ITT, and their combinations. At that time, Western Electric in particular was something of an unknown factor in the business of international competition. We had been apprehensive of the amount of muscle they could muster behind their proposal, and also aware of the fact, about a year earlier, that they had landed the contract for building the Saudi Arabian long-distance microwave network

About ten days before, Janet and I had spent a week in Costa Rica, taking part in an international telecom seminar. Over breakfast one morning, in conversation with a couple of representatives from Bell Labs and Western Electric, Janet had happened to mention that we were taking this opportunity to have a short holiday, since I had missed mine during the summer. She complained about the uncertainties for the wife of a chap working for Ericsson. This year, her husband had been fooling around with some hair-brained project for Saudi Arabia. It turned out that the fellow from Western Electric had had no holiday either, for the same reason. He had been tied up by their Saudi Arabia proposal! We had an interesting if somewhat guarded conversation, but Hank did suggest that, since he and his team had put together an unbeatable proposal, my summer had been wasted. I readily admit that privately I felt he might be right.

When I visited Teléfonos de México later that morning I was able to tell them the news, and the reference list I presented to support my AXE discussion had now become quite respectable.

In Stockholm that day, everybody was served a generous slice of cream pastry for lunch, with the compliments of Mr Svedberg and his thanks for a job well done. It still annoys me that I missed that celebration.

But it was only the very first part of the job. They were days of mixed feelings. There was, of course, joy and contentment at having beaten the competition. But there was also a certain element of uncertainty and some trepidation. We had landed this very large piece of business. How, in the name of Alexander Graham Bell, we asked ourselves, were we to carry it through?

But by the time I got back to Stockholm, things had begun moving, and fast. During the preceding month, it had become clear, although I was certainly not aware of it, that there was some chance that our consortium would be selected, and naturally certain preparations had already been set in motion.

The announcement of December 13th, 1977 was conclusive and binding (it meant among other things that we had to have certain exchanges in service within one year, by December 13th, 1978), but the actual contract document was not to be signed until January 25th, 1978. During the interim period, a story appeared in *Newsweek*, hinting that negotiations were still going on and that Western Electric would be the winner. The story was quickly taken up by the Swedish press. They must have spent a fortune on telephone calls to Riyadh trying to get details. I was one of the people in Stockholm to get called and though I had just returned from Riyadh, on January 20th, they refused to believe me when I explained how the contract discussions were proceeding and how the final text was being processed from English into legal Arabic and back to English.

Later, there was a good deal of speculation, even within the company, whether we had the capability and the resources to fulfil this large undertaking. The prize went to an interview with ITT's chief executive, Lyman Hamilton, which appeared in the Swedish business magazine *Veckans Affarer*. Mr Hamilton stated in no uncertain terms that Ericsson (Philips and Bell Canada were hardly referred to at all) had pulled off a monumental confidence trick on the Saudis; that Ericsson planned to deliver obsolete equipment, since it had no modern system; that the pricing terms of the contract were misleading, and that nobody, least of all Ericsson, could meet the in-service dates. He felt confident that Ericsson would be out of Saudi Arabia within a year. Mr Hamilton is out of ITT now, and Ericsson is still in Saudi Arabia.

Perhaps Mr Hamilton's dire forecasts helped to spur us on, and gave us an additional incentive to work hard. For there was much work to be done.

Industrialisation

The contract for Saudi Arabia had a profound influence.

It was very significant financially, of course—a huge piece of business.

But what it meant above all was an opportunity to take AXE quickly through the next stage of development—what is often called the industrialisation process. From being produced on a relatively small scale, during which we could improvise, carry out makeshift corrections, and generally help things along by manual means, AXE now had to be moved efficiently through the various stages of mass production—on the drawing board, in laboratories, in testing, in the factories, and in installation. It was being forced to grow up suddenly.

It also involved a major change in our production techniques. Our factories had been producing electromechanical systems almost exclusively, and now we were committed to electronics.

A temporary task-force was set up under Stig Larsson to review, plan and implement full-flow production of electronic switching equipment through the Ericsson organisation. We called it the INDAX (INDustrialisation of AXE) project.

A few examples of what it entailed are worth mentioning.

A function in AXE is implemented in hardware and software. The circuits for the hardware are designed onto a printed circuit board or sometimes several boards. At the time, we were turning increasingly to sophisticated support systems, including, for hardware designers, CAD (computer-aided design) systems.

CAD helps the engineer to design wiring patterns, for example, and lay them out on the boards, but however sophisticated CAD systems become, there's still a stage at which the manufacturing department checks the design in great detail to ensure it can be produced on the machines available, and can be tested by automatic test equipment.

This stage of preparation for production (which also involves extensive software testing and the production of test programmes) used to take anything from four weeks to three months.

We set the requirements at two weeks maximum—and it worked.

Another example is installation. In the days of crossbar, it used to be reckoned that it took 12 to 14 months to install and test one 10,000-line exchange. For Saudi Arabia, we had just 12 months to make design modifications, to have them verified and tested, and to manufacture, ship, install and test *three* large exchanges.

Installation not only took a long time, but was also very costly, since it involved keeping a number of people on site throughout the period of installation and testing.

The solution had to be to test the equipment before it left the factory in Stockholm. This had been indicated as a requirement in the original specification for AXE, and had been thought about a good deal. My boss, Hans Flinck, had been particularly active in trying to convince people that this was the only way to go. And Saudi made us go—very quickly!

A cable factory near the Stockholm headquarters was being converted to an electronic pilot plant, and a pre-installation plant was set up next to it. This is supplied (it still exists) with permanent racks into which the complete exchange to be tested is installed. It's a simple-enough operation, since all the cabling is plug-connected. The equipment is then tested in an environment which simulates the actual environment as far as possible. After testing, the magazines are disconnected, packed, and shipped directly to the final site.

Meanwhile, on site, the racks to receive the magazines have been built. The magazines can be put into position as soon as they have been unpacked, and final testing can begin within a few days.

We used air freight to Saudi Arabia, and cut the time to get a new exchange into operation down to two months.

The INDAX project was made up of some 60 different projects like these. When they were all implemented, we had a truly smooth and efficient production process.

Going places

By 1977, a large part of X Division had become entirely involved with AXE.

Transferring the new system from Ellemtel to Ericsson was a big and complicated business. We had realised from the first that this would be so, and the original small group of Ericsson people who were involved had made careful plans for handling it. It could be seen as an example of transfer of technology on a large scale, from the system design group at Ellemtel to production by Ericsson, engineering at Ericsson, and marketing by Ericsson.

The original system design that went into the pilot exchange at Södertälje was a basis for the continuing development of a working product. This continuous development was to be carried on by Ellemtel. Ericsson's technical departments would look after the design of the specific variations needed for different markets—what we call market adaptations. Ericsson would also undertake the design of some new subsystems.

The most important product of a design group is documentation. The pilot exchange at Södertälje was a triumphant verification of a major piece of engineering. As such, it was of the highest importance. But to maintain the build-up of a system family to be marketed world-wide, the key element is the wisdom that is accumulated and transmitted in documentation.

For SPC telephone switching, the documentation is extensive, varied, and complex. It can be divided into *production documentation*, which describes how to build and test the hardware and software of the equipment; and *operational documentation*, which describes how the system functions, and how it is maintained and administered.

There is also documentation for the training of our own and our customers' staffs; and, on a different level, the documentation needed for marketing.

Any SPC system also makes great use of support systems, or handling aids. These are very largely computer-based—such things as the programming system, CAD equipment for hardware design, or test equipment for use in production and installation.

In the original requirement specification, we had paid great attention to documentation. Our experience with our earlier systems had taught us that handling a great volume of documentation could become a grave problem. We suggested computer aids just to keep track of documentation; we suggested that the operational documentation should be designed and structured in such a way that it could also be used for training; and we had started looking at ways of storing documentation to make it more manageable—using microfilm, for example, and, naturally, electronic media.

Documentation and people combined to form the nucleus of the in-house engineering force that now came into being. We had a group of Ericsson engineers working on AXE at Ellemtel, who gradually transferred back to new engineering departments at Ericsson, bringing their experience with them.

From the very first order, market adaptations were needed. For Turku, for France (where Thomson rapidly built up their own engineering group), for Saudi Arabia, and for Australia, the work was done at Ericsson.

Meanwhile, we started to receive trainees from our customers. The training department began to develop courses for different categories of staff, including, of course, our own. Maintenance, software design, planning, traffic dimensioning—these and other technical courses got going. Several groups of our sales engineers took part, who in turn started up training programmes for our marketing staff and prepared the marketing documentation.

By 1977, we were fully committed to AXE. We had realistic plans for the development of a complete system family for virtually all the switching needs of a telephone Administration. We were well on the way to a competent, enthusiastic force of people for the engineering, production, and marketing of AXE.

At last, we could abandon our previous careful balancing act of presenting a product programme of three different SPC systems—ARE, AKE, AXE—and put all our resources into AXE and the digital network.

We were still selling small quantities of the other two systems, and even some crossbar, but our main thrust was now with AXE. We had reached cruising altitude.

JUMPING ON THE DIGITAL BANDWAGON

In 1977, the list of competitive products for telephone exchanges looked much the same as it had four years earlier. Many of the systems listed then had been offered to the market, had become operational, and were in production—though not yet on a large scale. A few systems had fallen by the wayside and were no longer on offer.

A year later, however, in 1978, the list had changed dramatically. The benefits of the digital approach, and the successes of CIT-Alcatel, Ericsson, and Northern Telecom of Canada, were becoming obvious. Our competitors had all gone back to their drawing-boards. They were putting all they had into bringing out new systems, digital systems. And as these new products began to turn up in technical papers, in symposia, in exhibitions, and in tender proposals the international menu of offerings to telephone Administrations acquired a completely new flavour, and the analogue systems began to disappear.

If we draw up a list of switching systems on offer from around 1978, we find this new picture. There are no longer two or three of us offering digital systems—there are nearly a dozen. The competition looks as though it is catching up, and our competitive edge looks as though it is threatened. Obviously, we tried to follow what was going on very closely. The new list of competing systems on the market looked something like this:

	Earlier, analogue systems	Later, digital systems
Western Electric (US)	No 1 ESS	NO 5 ESS
ITT (US, Belgium, etc)	Metaconta	System 1240
GTE (US)	1EAX, 2EAX,	5EAX
Northern Telecom (Canada)	SP 1	DMS 10, 100
Siemens (Germany)	ESK, EWS-A	EWS-D
Philips (the Netherlands)	PRX	PRX-D
Thomson-CSF (France)	(AXE, analogue)	MT 20, MT 25
Nippon Electric (Japan)	D-10	NEAX 61
Fujitsu (Japan)	D-10	Fetex
Hitachi (Japan)	D-10	HDX 10
OKI (Japan)	D-10	KB270
British Telecom (Plessey, GEC, STC)	TXE 4	System X
CIT-Alcatel (France)	E 10*	E 10
NOKIA (Finland)	—	DX 200
Stromberg-Carlson (US)	—	Century

*Note: E10 was digital.

All the systems listed are for local exchanges. Other acronyms might have been included for transit applications, tandems, and so on.

All the systems listed are designed as components in the *digital* network. This means that they all offered digital group switches.

By the late '70s, building a digital group switch was a fairly straightforward matter. The key components, memory chips, were available, and not too costly. CIT-Alcatel had, after all, been producing such switches since the beginning of the decade.

But the subscriber switch is a different matter. Component availability is certainly crucial, but there is also a choice of several different principles for the design of a subscriber switch.

The drive to produce a digital subscriber switch has different roots at different stages of development.

If we ignore data transmission, there's no great importance attached to a digital subscriber switch. What matters is a subscriber switch that can be "remoted" from the main exchange—a switch that can be physically positioned in the network near subscriber concentrations. Such units are also called subscriber concentrators. A concentrator may be designed using any technology we choose, as long as it is cost-effective, has low power-consumption, and requires little space.

In the longer term, however, as we move towards the integrated services digital network, ISDN, in which the subscriber lines from the exchange to the subscriber premises will carry digital signals (speech, data and text in binary form), the subscriber switch has greater requirements to meet. Even now, however, it doesn't *have* to be digital, although many manufacturers have chosen digital technology.

Ericsson is pursuing the digital approach, although the versions of AXE shipped up to 1982 have incorporated analogue subscriber switches. Other systems use electronic crosspoints, pulse amplitude modulation (PAM), or combined techniques.

Some members of the competition have chosen to talk in very disparaging terms of AXE and other successful systems as "hybrids", and not purely digital. It depends on your standpoint. You may believe that it is the choice of technology that decides how good a system is, or you may judge according to its features, its costs, its quality, or its other characteristics. Certainly, you must give some weight to whether the system actually exists. Many of the products on the list above have *still* to be proven in operation.

THE END OF THE FIRST TEN YEARS

By the end of 1977, we had contracts for some 480,000 lines of AXE, for 31 exchanges, in seven countries. AXE had become well known in the world of telecommunications. The Saudi tender competition was one of several that characterised the end of the decade, and we submitted proposals for most of them. With each proposal, more of our own people became familiar with the system in detail, and more of us became expert in analysing the requirements of different customers, and tailoring the system to the specifics of different networks. Our battery of aids was growing—computer systems to simplify, for example, the specification process, which can be tedious and time-consuming when tackled manually, and in which the human factor may introduce costly errors.

Worldwide, acceptance of the concept of the digital network, its immediate economic advantages, and its value as a carrier of new services in the future was also growing. As the digital revolution swept the world, we were there. We were ready with the right product, at the right time, and at the right price—and we were one of the few manufacturers to be in this happy position. There were plenty of international tenders about, and it was up to us to make the most of the situation.

But the competition was active, too. Although many of our competitors didn't yet have comparable products, they certainly made us work for our money. A good product and an acceptable price do not win contracts on their own. Thinking back over the many competitive situations we found ourselves in, we can list a number of factors which illustrate the difficulties of the job—made more difficult by the world's economic ills and the strained situation in the telecommunications industry.

Increasingly, finance has become crucial. The manufacturer must often be able to arrange—or even himself supply—the loans to finance the expansion of the customer's network. Soft money can be decisive in getting a contract, more important than the type and quality of the exchange equipment. Since a national telecommunications industry is always a vital part of a country's economy, many governments support their own industry's efforts to export, sometimes by underwriting credit offered, sometimes by subsidy through favourable interest terms. We've touched elsewhere on the World Bank rule that contracts must go to the lowest bidder, and on its consequences. It also irritates us when a country offers a package deal for arms or other attractive goodies, with telecommunications included as a condition for a deal. As Swedes, we can't match these deals, and it puts us at a disadvantage.

Another important factor is the vendor's ability to provide service at all levels, and after, as well as before, a sale. Vendor-customer relationships are long-term; if they are cordial, there is an excellent chance of follow-up orders for network extensions over many years.

We've said already that a prime manufacturer must be able and willing to establish local manufacture in customer countries. At one time, Ericsson's largest factory was in Tsarist Russia, and we also had large plants in Britain and the US, amongst other places. This, of course, was before the first world war, but the tradition has continued. Ericsson crossbar is, or was, manufactured in 22 countries, in our own plants or under license. This has given us a solid manufacturing base around the world, into which we are gradually introducing the local manufacture of AXE. As the world becomes aware of the importance of electronics, there is a new trend towards local plant ownership. In some countries, there is a positive requirement that production facilities must be locally owned, and Ericsson has had to give up majority ownership of some companies. In these circumstances, management is under great pressure to perform well. Adapting to change is a prerequisite for success in the business—a necessity for survival.

So professionalism on its own is no guarantee of success, but being entirely professional is important. The nicest compliment we've ever had was from a customer who was himself very professional: "You are professional—at all levels". We have to be, or we shouldn't be here today.

Certainly by 1980 AXE was a fully-fledged international system. At the turn of the decade, we were involved in projects all over the world.

A great part of my heart is still in Mexico. I've been fortunate enough to have many opportunities of visiting the country from time to time, and I always feel at home there—though Mexico City has become a trying place, blanketed by fumes and with traffic at a stand-still for large parts of the day.

In 1978, I joined a group from the local Ericsson company to visit Tijuana, a town on the border of California. Tijuana, Mexicali, Enseñada and the surrounding country had been served by an independent telephone company, but this had been taken over by Telmex. It is still run as a separate entity, called Teléfonos del Noroeste, Telnor.

Several members of its management are old friends and colleagues from my days as an engineer with Telmex. Telnor had committed itself to the local government to sort out the service in the area quickly. We had been asked to make a proposal for the switching equipment, while Telnor itself got down to refurbishing the cable installation and other parts of the service. Plans were drawn up for a digital network, based on a combined local/transit exchange (which would also handle cross-border traffic to the US) and several new local exchanges with remote units serving outlying subscriber concentrations. The first local AXE went into service in 1980 in Tijuana. Today, a further 25,000 lines are in service, and the first digital concentrator was cut into service in 1982.

Telmex, meanwhile, was undertaking the huge job of

specifying and selecting a new switching system. ITT and Ericsson have been the suppliers to Mexico since the beginning of the century, and we both have large factories in Mexico. When the specifications were issued, they called for a digital SPC system to be manufactured in Mexico. Seven companies submitted proposals. The finalists were ITT and Ericsson, and in the last round Telmex decided that ITT's System 12 would be used for 75 percent of the installations and AXE for 25 percent. This was a blow, especially as ITT's 75 percent excluded us from Mexico City, where we'd been the main supplier for years. However, this was a long-term agreement. Contracts for the future were contingent on each manufacturer's cutting in a pilot exchange (maqueta) in Mexico City by June 1981, to be used by Telmex for training and further evaluation. We had just a year to do it, but with a crash programme we got the AXE exchange into service on schedule. The System 12 machine is next door in the same building. As this book is being written (June 1984), the System 12 exchange is not yet in service, though we understand that it will be later this year.

As it happens, Telmex has had traffic-handling problems in Mexico City for some years, and the point came where it needed certain tandem exchanges in a hurry to relieve the situation. We have been able to help out with a number of AXE exchanges, so we feel happier about the whole situation.

In 1979/80, we made strenuous efforts to get the contracts for Egypt. Ericsson crossbar exchanges have been manufactured in Egypt under license for many years, but the Administration was now determined to modernise the network with stored program control. At the same time, it proposed to update the network, which was severely run down. The cable network was in particularly bad shape, owing to Cairo's unfavourable sub-soil conditions. We made several presentations, and finally a full proposal. We were given to understand that our proposal was the least costly, but this was a case where financing was the deciding factor. A Franco-German consortium made an offer of long-term, low-interest credit which was unbeatable. The equipment is to be supplied by Siemens and Thomson. CIT-Alcatel has sold some E10 to Egypt, and Ericsson some AXE and ARE.

In Malaysia, we got a few straight orders for AXE, then lost the first large contract to Nippon Electric. Malaysia, however, had decided to have two suppliers to avoid monopoly, and the second contract went to Ericsson. Both these contracts are ten-year, long-term agreements for about a million lines each. Similarly, in Thailand, the Japanese got the large orders for exchanges in Bangkok, but a second set of orders went to Ericsson. Malaysia and Thailand are old Ericsson markets, and it was a relief to be back after nearly being squeezed out.

The dynamic industrial development of Korea is familiar to everybody. Korea attracted the attention of every manufacturer in the world when, towards the end of the 1970s, it announced its intention to go for a new system to replace its existing Siemens EMD system, manufactured under license. (There was also a license agreement with ITT for Metaconta, and the first Metaconta exchanges were nearing completion.) I made my first trip to Seoul to present AXE in 1978. The occasion was a one-day seminar arranged by the Korean PTT, during which each of eight manufacturers had been asked to make a one-hour presentation. This was only the beginning of a long process of evaluation. Eventually, Western Electric won the first round with a contract for the large metropolitan exchanges—some 500,000 lines in all of the No 1 ESS system. So the Koreans chose an analogue system, but in the second round we managed to get the order for a number of transit switches in major Korean cities. A third contract, which covers a programme for the towns outside Seoul, and which includes local manufacture, was awarded to Ericsson in 1983. I had a personal interest in the outcome of these negotiations, since my son, Peter, now an Ericsson engineer, spent a lot of time there negotiating and planning the setting up of production and supply facilities.

In all this, my own role has been a minor one. I have been one of the preparers of the ground, going in early to talk to the staffs of the Administrations and our own representatives. This has often involved presentations to the top managements of the telecommunications companies and their boards. It's a sign of the times that the decision on the choice of a telephone system now very often lands high up in the organisation. In the early days, it was strictly a matter for the engineering departments. I have found myself talking to people who are not engineers—politicians, lawyers, accountants. But then, the conclusive arguments for SPC and digital switching and transmission are not really technical. The engineers are fascinated by the technology, and enjoy discussing it, but management is interested in money. If technology can help save money on plant investment, and operation and maintenance, we shall have the attention of management. So my colleagues and I began to gear our presentations to explaining and discussing the economic blessings of the new technology. Our papers for seminars, advertising, and exhibitions have followed suit.

And this is important. There is still a tendency in the telecommunications community to focus over-intently on technical detail, and to maintain, for instance, that a new system which employs a revolutionary new component must be better than its predecessors. If it reduces the cost, and if quality, performance, traffic-handling, maintainability and so on are not adversely affected, fine, you have a better system. But technology for technology's sake is rubbish.

Success on record—II

BY July 1978, the reference list had grown longer, and to get it onto two pages of overhead film we had to resort to smaller type.

July 1978

Country	Local Exchanges	Multiple Capacity In Service	Multiple Capacity On Order
Finland	Turku	4 000	9 000
France	Argenteuil		12 000
	Chalon s/Marne		7 000
	Chartres		12 000
	Joué les Tours		9 000
	Longwy		8 000
	Metz		12 000
	Nancy		12 000
	Nantes		10 000
	Nevers		8 000
	Orleans		13 000
	Strasbourg		6 000
	Versailles		8 000
Kuwait	Ardiyah		10 000
	Mina Abdullah		10 000
	Mishrif		10 000
	TEC-C		20 000
	Hawalli		10 000
	Ras-Salmiah		10 000

July 1978 *(Cont.)*

Country	Local Exchanges	Multiple Capacity In Service	Multiple Capacity On Order
Panama	Juan Diaz CO 20		3 000
Saudi Arabia	Al Malaz 2		20 000
	Al Ulaiya 2		20 000
	Medina Road 1		17 000
	Medina Road 2		18 000
	Mecca 1B		15 000
	Mecca 3B		15 000
	Mecca 4A		28 000
	Khdayriah		20 000
	Dammam 1C		8 000
Spain	Atocha		10 000
Sweden	Aspudden		34 000
	Boras		40 000
	Gavle		36 000
	Jonkoping		10 000
	Karlstad		10 000
	Masthuggel		38 000
	Savedalen		10 000
	Södertälje	3 000	32 000
	Tumba		10 000
	Vasteras		10 000
	Ostermalm		10 000
Venezuela	Caracas		5 000
Yugoslavia	Vrapce		10 000
Total		7 000	625 000
		632 000	

End of table.

July 1978 *(Cont.)*

Country	Transit Exchanges	Multiple Capacity In Service	Multiple Capacity On Order
Denmark	Alborg		9 000
	Kolding		9 000
Finland	Turku	480	
Malagasy	Tananarive		4 000
Saudi Arabia	Jeddah SC		9 600
	Mecca PC		9 200
	Dammam SC		6 200
	Riyadh SC		8 600
	Buraydah PC		5 300
	Abha PC		5 800
	Medina PC		8 200
Total		480	74 900
		75 380	

We had now added the Saudi Arabia contract with a series of local and transit exchanges; we had had a second order for three new exchanges from Kuwait; we had two new markets, Venezuela and Panama; and the digital tandem in Turku, Finland was now in service.

The following is from six months later, December 1978, (on pages 149 and 150).

December 1978

Country	Number of Exchanges		Multiple Capacity in Service		Multiple Capacity on Order	
	Local	Transit	Local	Transit	Local	Transit
Australia						
Colombia	11				100 140	
Denmark		3				24 600
	1				10 000	
Finland	1	1	4 000	480	9 000	
France	41				377 000	
Italy		1				3 800
Kuwait	6				70 000	
Madagascar	1				20 000	
Netherlands						
Norway	2				25 000	
Panama	3				10 000	
Saudi Arabia	9	7	20 000	10 500	141 000	42 300
Spain	1				10 000	
Sweden	11		3 000		240 000	
	2				20 000	

(Cont.)

December 1978 *(Cont.)*

Country	Number of Exchanges		Multiple Capacity in Service		Multiple Capacity on Order	
	Local	Transit	Local	Transit	Local	Transit
Venezuela	1				5 000	
Yugoslavia	1				10 000	
Zaire	1				6 200	
Total	92	12	27 000	10 980	1 053 340	70 700

End of table.

By now, we had decided to skip a lot of detail and were no longer naming each individual exchange. The first batch of Saudi exchanges was in service, both transit and local. The Australian Administration had made its decision in our favor—AXE had been chosen as the standard system for the Australian network. The Dutch PTT had chosen AXE as one of two standard systems, alongside the Philips PRX. The list also includes the so-called MTX exchanges, for Norway, Finland and Sweden. An MTX exchange is an AXE transit which includes a subsystem for connecting radio base stations for mobile telephone service (also called cellular radio). These five MTXs were to form the backbone and supply the processing power for the Nordic Mobile Telephone Network. Thomson CSF was marketing AXE in accordance with its licence agreement with L M Ericsson. Apart from the French orders, we find two markets, Madagascar and Zaire, ordering AXE from Thomson.

We now had well over a million exchange lines on order and in service, not counting Australia and the Netherlands. A very respectable result indeed, we thought—reached in only about two years of active marketing. And we continued to be very active. These were the years when many Administrations were inviting tender proposals and making decisions on new systems. We tried to spread the gospel of the digital network wherever we went, and more and more telephone companies were now specifying digital exchanges. This gave us an advantage—we were one of only three manufacturers who had digital switches in operation. AXE had other advantages too, modular software, for example. And we also had a most sophisticated operation and maintenance system. In addition, of course, we had begun to promote mobile telephony. We were still maintaining our two-year lead over the competition.

Just over a year later, in January 1980, AXE had reached over two million lines on order or in service for 23 countries and a total of 211 exchanges. The number of lines in service was big enough now to provide important feedback in the form of service records. And the service records were excellent.

January 1980

Country	Number of Exchanges		Multiple Capacity On Order		Multiple Capacity In Service	
	Local	Transit	Local	Transit	Local	Transit
Argentina	1		7 000			
Australia	1		4 000			
Bahrain	1	1	10 000	6 000		
Brazil	5		50 800			
Colombia	24	1	230 000	12 328		
Denmark	2	4	6 000	26 000		
	1		10 000*			
Finland	1	1	19 000	1 440	4 000	480
	1		10 000*			
France	76		659 584		12 900**	
Italy		4		7 680		4 800
Kuwait	6		110 000			
Madagascar	1		20 000**			
Malaysia	2		30 000****			
Mexico	10		25 500			
Netherlands	16		42 496			
Norway	2		26 000*			
Panama	3		10 000			
Saudi Arabia	15	9	93 000	24 480	90 000	42 240
Spain	3		30 000			

(Cont.)

January 1980 *(Cont.)*

Country	Number of Exchanges		Multiple Capacity On Order		Multiple Capacity In Service	
	Local	Transit	Local	Transit	Local	Transit
Sweden	12		242 000		3 000	
	8		138 000****			
	2		20 000*			
United Arab Emirates	5	1	43 400	3 000		
Venezuela	1				5 000	
Yugoslavia	12	2	75 000	5 135***		
Total	211	23	1 910 780	86 063	114 900	47 520

* System for mobile subscribers
** Manufacture by Thomson CSF
*** Manufacture by Nikola Tesla
**** Manufacture by TELI

We may notice a new group for Sweden. This was the first batch of exchanges ordered from the Swedish Administration's own factories, TELI. The exchanges for Mexico at this stage were for Teléfonos del Noroeste.

There were tender competitions in which we lost. Thomson CSF was now marketing its MT 20/25 system and being quite successful in getting orders. Some of these customers have suffered from long delays in getting the MT system operational. And around this time Thomson stopped marketing the AXE system. As a matter of fact, there are very few references to the AXEs installed and ordered for France, and the Zaire order for AXE is no longer on the list. Thomson had successfully proposed to the Zaire people that their AXE order should be converted to MT. After all, the MT system is digital, which the Thomson AXE is not.

The Japanese companies, Nippon Electric and Fujitsu, were successful in getting orders in Argentina, Singapore and Hongkong, among other markets.

CIT-Alcatel was clocking up orders at a good rate, and Siemens had begun marketing its digital system, EWS-D. The Bundespost had made a complete turn-around in 1977 and decided that the German network should be digital. Northern Telecom of Canada had a message very similar to ours as far as the digital network was concerned—about this time it made a break-in on the US market and began supplying the Bell System. Of the US companies, Western Electric is the giant, of course. It was now promoting the No. 5 ESS, and had also for some time been supplying the No. 4 ESS for toll offices (transit exchanges). Western Electric made some attempts to get into the export business. It was one of the contenders in Saudi Arabia. It won the first round in the Republic of Korea to supply the large local exchanges, and also picked up some business in Taiwan. GTE, the second largest American company, was also trying to export, but the small company Vidar, now a division of TRW, who had been pioneers in building a digital system for rural areas and small towns, was having difficulty finding a sufficiently large market. Vidar has since gone out of business.

Competition was extremely tough; there were many large tender competitions; and by now all the competitors were offering ditigal systems. Even so we estimated that we still had the two-year technological year lead over most of our competitors, mainly because of the difficulties several of them were having in making their new systems operational.

CHAPTER 5: WHERE DO WE GO FROM HERE?

GROWING UP

Södertälje was a pilot installation in which we put the new system to the test of handling live traffic. But the Södertälje version of AXE was far from a final product. It was really just the first stage in the system-development process which was to continue, and which still continues today.

We have seen how the first installation of a digital version came into service a little over a year later, in the Turku tandem exchange. Then in Saudi Arabia, beginning in 1978, we have the first installation with big local exchanges carrying high traffic loads. These exchanges had an enlarged digital group selector, while the subscriber stages were, of course, analogue. Also in Saudi Arabia, and also in 1978, we got the first large tandems going. With digital tandems and big local exchanges with digital group selectors, we were in effect building a basic digital network. We could see how the combination of digital switching and digital transmission over PCM systems was working in practise.

In Saudi Arabia we also had centralised operation and maintenance working on a large scale for the first time. One of the characteristics of AXE is that a very large proportion of the applications software (the APT programs, you will recall) is devoted to handling administrative functions—about 70% of it provides features for operation and maintenance, supervision, and so on. There were two guiding principles in the design. The first was to provide a very full menu of functions. The second was to do this as an integral part of the system—each exchange should be fully self-contained, not requiring such additional test gear as monitoring equipment, test call generators, subscriber line measuring apparatus, and so on. This concept is not new, but we used the experience of the Swedish Administration to take it to its complete, logical implementation.

In the old days, exchange maintenance was based to a very high degree on routine testing. It might be a rule, for example, to check through all the trunks in the exchange every day, or to run through all the registers to be certain that they were functioning correctly. When crossbar arrived in the '50s, with its far lower incidence of faults, these routine tests yielded meager results. There just weren't enough faults to make all the testing worthwhile. It began to seem logical to leave out most of the testing, and do nothing until a fault actually occurred. It was a different approach to design—the exchange must itself be able to signal when something was wrong, and we started to incorporate circuitry for this purpose in our designing. The concept is called CCM, controlled corrective maintenance, and it became pretty general in the '60s and '70s.

With the greater logical power available through the processor software of AXE, we took the CCM principle a giant stride further. *All* routine testing became unnecessary. The exchange supervised and monitored itself, and signalled to the maintenance staff when something was wrong.

Maintenance people for SPC systems are highly skilled. The equipment is complex, and there's a lot of it. For staff with a basic knowledge of electronics and computer technology, the formal training does not take long. For someone with some experience of telephony, a few months is normally plenty. But modern electronic equipment doesn't develop faults very often, so anyone put in charge of an AXE exchange has not much to do. However well trained he may be, he will have little opportunity to maintain his skills at their peak. Without the need to use his skills, he may lose them. Even worse, he may get bored and start playing with the equipment, which will cause trouble if he happens to muck up the software.

We sometimes illustrate the staffing requirements of an AXE exchange by explaining that all an exchange calls for is a man and a dog. The dog is there to see that the man doesn't touch the equipment. The man is there to feed the dog.

The fact remains that operations and maintenance are a different proposition today, with electronic SPC exchanges, and call for a different approach. The answer is centralisation. Exchanges are supervised remotely, and the necessary staff are concentrated at one or a few central points in the network. Each exchange looks after itself, monitoring the live traffic that passes through. Only if a fault occurs, or if the automatic statistics indicate that there is one, does a maintenance person set up some man/machine communication and go into a diagnostic program. Hardware faults will still call for a site visit, to replace something (normally a printed circuit board), but these visits are hardly emergencies. The system sees to it that traffic is not disturbed. Remember—anything vital is duplicated; a single fault will not disrupt the service or cause deterioration of the traffic-handling capacity.

In Saudi Arabia, the network contains AXE, ARE and Philips PRX exchanges. Operations and maintenance computers are connected to all the exchanges over data links, and act as communications centers, directing the data links to the work-stations which correspond to the different administrative tasks to be performed. Such work-stations are normally in the central maintenance office, just described. There may be another centre for traffic metering, from which are programmed a series of traffic measurements over the year. Such measurements provide important statistics for planning new plant, network engineering, and traffic-routing directives. A third centre may be used for customer services. This is where you order a new telephone, and where your bills are paid. If, entirely by oversight, you fail to pay your bill, you will eventually find your phone has been disconnected. When this happens, you will, of course, find the money and rush in to pay your bill. Normal procedure has then been that it takes about 48 hours (and a lot of paperwork for the telephone company) for you to be reconnected. With AXE, it takes a few seconds. Someone in the service centre punches a simple command plus your telephone number into a terminal, and you're back in business.

One further type of maintenance centre is worth mentioning: the repair service centre. This is where you call when your phone is out of order. From here, your line will be checked, and, if necessary, a repairman will be sent to find the trouble and put it right. Today's repair centre is equipped with only a terminal. All the measuring and testing equipment that used to be the rule is integrated into the exchange. It is the exchange which tests the line and decides whether the cable is too damp, or checks your dial or key set. Depending on the organisation of the telephone company, there may well be further centres, all accessing the exchanges and the rest of the network components through the operations and maintenance computer. It is called AOM 101.

The development of the AOM system is just one example of the extensive and continuing programme of building up AXE as a strong system. This programme has three distinct areas.

New applications

New markets mean new applications. In its first applications, AXE was a straightforward analogue local exchange. The digital tandem came next, used as an intermediate switch in multi-exchange city networks. Over the years, applications have multiplied, each requiring fresh design work, principally in software, but quite often in hardware, too. The first application of AXE to a transit exchange was analogue, on paper, and used a four-pole reed relay. As soon as the digital switch came in, we scrapped this design, and in fact all transit exchanges have been digital. An international transit calls for a lot of special functions to handle complicated international signalling functions, assistance operators, and intricate revenue-calculations.

As we saw when we first looked at the network, the remote subscriber switch, or concentrator, is an important component. We made the first ones analogue, but, of course, the remote switch really comes into its own in its digital version—small, absorbing little power, ready to fit into the ISDN of the future. We got the digital version into service in several countries during 1982.

New functions

Most markets come up with their own more or less unique requirements. Sometimes these are trivial—signalling interchanges, say, with a particular PABX. However trivial they are, they call for some engineering effort, and the writing of new software. Sometimes they are far from trivial, like the requirements which derive from an Administration's particular administrative routines, which might include the production of charging data to be handled in an existing computer system.

Now that we have sold AXE to over 70 telephone Administrations, in over 50 different countries, and met all the different requirements for specific functions, we have a very extensive library of functions to call on. This is a most valuable investment on which we can draw for many years. Because AXE is modular—functionally modular—we can select the appropriate function modules for any combination of functions required, and put them together, undertaking new design only for completely new requirements.

Mobile telephony is a special case, falling between the new applications and new functions categories. In the early 1970s, the Administrations of Sweden, Norway, Denmark and Finland got together to write a joint specification for a common international mobile telephone system. It was natural to look at AXE as the nodal point in such a system, a node which would be the interface with the national network, but which would also provide the system's intelligence. The system is now in service in the four countries providing an integrated mobile telephone service. There is a uniform numbering scheme, and provided your car is in Scandinavia, a call can reach you without knowing where you are. The system keeps track of you by a system of identification signals, and makes sure that you are always receiving from the radio base station which will give you the best transmission. The system can incorporate the small-cell technique, which is specially designed to cover high-density areas, such as cities, with the minimum of radio frequencies. Mobile telephone systems, all with AXE switches, have now also been sold to the United Kingdom, Ireland, the Netherlands, Saudi Arabia, Oman, Spain, Malaysia, Tunisia, and Chicago, Buffalo and Detroit in the US.

New technology

Perhaps the most dramatic effect of the entry of electronics into switching is the rate at which components change and develop. In some ways, this presents the industry with a dilemma—this year's design may be obsolete next year. It's true that the functions don't change, as a rule, but component development usually cuts the cost of supplying a given function, and manufacturers are naturally expected to supply their system in the most cost-effective form. When we wrote the requirement specification of AXE, we were fully aware that this would happen, and again it is the functional modularity of AXE which allows our customers to benefit from advances in technology. Each module is a "black box". Provided we keep the structure of the system stable, and maintain the standard interfaces, we can put anything we like into the boxes. It sounds simple and obvious, but at the time it was a genuinely new concept in switching.

Today, the AXE that comes out of the pre-installation plant *looks* nothing like the Södertälje exchange, apart from the racks with their protective covers. New technology has been introduced into most parts of the system as it has become available.

The most striking case is that of memory. The system started out with memory chips of 1000 bits per chip. Then we had a period with 4000-bit chips, then 16,000, and today the chips we are using are 64,000-bit. Processors and digital switches are both built to a large extent of memories, and the physical size of the exchange has consequently shrunk dramatically. It will continue to do so.

Processors provide another example. The original regional processor was a mini. Today, it's a micro. The CPU is also micro-based today. Its size has been cut dramatically, yet it has six times the traffic-handling power.

AXE is not static. It is continuously upgraded and improved. But—and this is the important point—the bulk of the software remains the same, and the architecture of the system is unchanged. AXE is not static, but it *is* exceptionally stable, and exceptionally well proven in just about every conceivable situation.

SHOWING OFF

Telecommunications exhibitions are good business—for the organisers. To us in the industry, they are a pain in the neck and a great expense. Of course, trade shows have been around for a long time—the USITA (now USTA) show in the US is an example. But in the mid '60s, maybe with the advent of electronics and stored program control, it became increasingly important to take part in shows with a large section of public switching systems. One reason, no doubt, was that the industry sensed the coming of tougher competition, and the emergence of an increasing number of national industries in the business of telephone switching. At that time, too, awareness was growing of the links to come between computer technology and telecommunications technology. Whatever the reasons, we found that we were coming under pressure to buy space in exhibition halls and put our equipment into these shows. We were reluctant; this was a new venture, and many of us, especially of the older generation, could see no earthly use in exposing ourselves like tarts in this cheap and commercial way. L M Ericsson had been around for close on a hundred years, we had always managed to maintain excellent relations with our customers around the world, and everyone that mattered knew us and our products. So why resort to cheap commercialism? For myself, I am still of two minds.

One of the important aspects of exhibitions is no doubt that they give us a good chance to see what the competition is doing, how far they have got with competing products, and what particular product characteristics they are promoting. There is always a good deal of fraternization; we are given (or steal) each other's printed material, and we try to chat up each other's hostesses. We have been to shows where we found most of the visitors were from the competing exhibitors—and surely there are more efficient, and cheaper, ways of exchanging information.

But some shows are better than others, and we do take part in several every year. Some exhibitions are more commercial than others. Some include seminars at which papers are presented from Administrations, the academic world, and the manufacturers. We have made a private ranking, identifying some seminars where we can be frankly commercial, and make our papers into straightforward product-promotion exercises; others where it is prudent to hide the propaganda under an umbrella of technology principles or research results; and a third category in which we present technical subjects seriously.

The AXE system was first presented to the public at the International Switching Symposium, ISS, in Munich in September 1974. As I've said elsewhere, it was a paper in a low key, discussing switching system requirements and some new design principles. Although ISS is considered the most serious, professional and prestigious gathering of the lot, the AXE paper that time caused very little comment. The sessions were dominated by Bell Labs, who reported extensively on their Number 1 ESS system. During the conference, we also visited the first installation of a Siemens EWS exchange in operation, and were duly impressed.

The Telecom exhibitions and conferences are sponsored by the ITU, and take place every four years in Geneva. We had a small stand there in 1975, and among the many products we showed was a rack of AXE. We were back in 1979 with a larger stand, devoting a lot of space to the digital network, and also exhibiting some new concepts for rural network building. Telecom in Geneva, in our private classification, is set down as the in-between type as regards commercial flavour, and in the paper I presented I discussed a series of new network principles—without once mentioning AXE. In this, I apparently distinguished myself from the other members of the panel, so that even the chairman, bless him, noticed and commented on the fact. We got a free plug, while maintaining a strictly technical profile.

Intelcom 77, in Atlanta, US, was organised by the Horizon House company along strictly commercial lines. For various reasons we decided to participate, one reason being that a large number of visitors from Latin America were expected. Only a few of the main competitors took part in the show but a lot of them showed up as visitors and took part in the seminars. We had Thomson CSF in the stand next to ours, and since at that time they were still showing AXE, the system got good exposure. Kjell Sandberg and I had a paper each in the conference and we certainly provoked a good deal of interest and many questions. By now we had two AXE exchanges in service, in Sweden and in Finland, and had the contracts for France and Australia, so people were waking up to the fact that Swedes might be worth taking account of and listening to. Both Kjell and I plugged AXE unblushingly. It was that sort of show. Intelcom 77 also marked the first time we tried the American way of life. Since then we've had a lot more of it one way or another, and we have now also had our first orders for

AXE from the US.

There are a host of other exhibitions, some every year, some every third or fourth, some as once-only occasions. Rio de Janeiro, Singapore, Bahrein, Paris, Nairobi and so on. G O Douglas has been in charge of organising most of them, and he is the perfect commissar. But the work to prepare and plan and build is immense, and the cost hurts. We have had conversations with some of our competitors, strictly on the Q.T., of course, about curbing these activities, or at least limiting the wild growth.

Shows and conferences are often also the venues for receptions. The large organisations, or sometimes the collective representatives of a country, invite everybody to an evening of drinks and a buffet. We have hosted several of these. At their best they provide a good opportunity for meeting our customers in a relaxed setting; at their worst, it's just a free-for-all. I recall two such memorable parties that we have hosted. One was during an ITU conference in Paris. We hired a boat on the Seine for the evening. There was Scandinavian and West Indian food and booze, Parisian accordion music for dancing, and beautiful views of Paris as we cruised along in the spring evening. It was a great success.

The other occasion was during the Miami exhibition in 1981. We had nearly 400 guests, mostly from Latin America. Latin American food and drink and, above all, Latin American music. I have never, before or after, experienced such go in a large party—the noise was deafening, but it was fun.

We get invited to, or crash, other parties, of course. Kjell Sandberg and I have a private scheme for increasing the competitors' costs on these occasions. We set about drinking as much as possible of the most expensive liquor; and we lay in quantities of caviar and paté de foie gras. We haven't yet managed to drive any of the competitors into liquidation by this scheme, but it does drive them to distraction, and it bolsters our morale to feel we are doing something positive about something, even in a small way.

To us the most interesting exhibition and series of seminars happened in January of 1979 when we took part in an exhibition arranged by the Swedish Export Council in Shanghai. About 20 Swedish companies took part. The Chinese were most helpful. Because of the obvious language difficulties they provided us with translators and what they called "technical explainers"; the latter looked after the stand and were excellent as guides and hosts. I enjoyed listening to the digital network principles being explained to group after group of attentive Chinese visitors. And we certainly had visitors in large numbers; there were long queues outside the exhibition hall every morning and afternoon. In parallel to the show, we ran a seminar for the people from the Shanghai branch of the Chinese PTT and for technical students. This was definitely a category one occasion, so we were strictly technical—Bengt Gunnar Magnusson, Inge Jonsson, Walter Widl and myself.

Shanghai in January was pretty cold, and there is no central heating (except in the hotel reserved for foreigners), but after a few days we got used to going to the opera in our winter overcoats and long underwear. G O Douglas was on stage for a conjuring performance, and performed well—though he never has been able to figure out how the pretty lady got out of the box he had nailed down. I have a photograph of a group of us eating lunch at the Shanghai airport restaurant, with chop-sticks, and with gloves on.

Janet came along on this trip, and we also visited Hong Kong, Singapore and Indonesia. (I had credit with her for months!) I have a number of Chinese paintings in my office—a reminder of a most enjoyable visit. And Ericsson has had a few orders for AXE for China.

TELERICA

"Telerica? Of course I've heard of it, but can you tell me just where it is, exactly?" The question came from a serious American lady, representing a serious telecommunications journal at the telecommunications exhibition in Miami in 1981. I had some trouble explaining.

Jeans also had a lot of trouble with Telerica. That is when we actually met and got to know each other, for better or for worse. This is the story.

We had been putting a lot of serious effort into an exploration of the effects of the digital network—its planning, installation, operation and maintenance. We had paid particular attention to the economics, using real situations from different countries. We had analysed the capital costs, comparing them with the traditional investment costs involved in building analogue networks, and we had also gone deeply into the economic implications of operation based on centralised concepts. We had built up quite a collection of digital case studies, and were discussing different methods of sharing them with potential customers. The first proposal was a series of advertisements, each dealing with a particular phase of digitalisation, or a particular sample application. I discussed this proposal with G O Douglas, who asked me to try to sketch out my ideas, and then called in Richard Jeans from London. A few days later, Richard arrived in Stockholm with his umbrella and a carton of cigarettes.

As I've mentioned in my introduction, Richard took a fairly dim view of my approach. This didn't exactly endear him to me, especially as I had spent a lot of time working out the details. Obviously, he needed the job badly, and he tried to make the most of explaining how the whole thing had to be reworked from the beginning, laying down the law about the basic principles of advertising, graphic design and copywriting. I didn't really believe much of it, but we needed extra capacity to get the thing done, and after I'd spent an inordinately long time explaining to Richard the most elementary things about digital switching (and indeed telephony) he went back to London to get to work.

While Richard was, I suppose, busy working out the best way to get the most money possible out of Ericsson, G O and I were discussing the case studies. At first, we thought of locating them in different real countries to provide the necessary working environment, but after some time we came to the conclusion that we needed a whole new country, a sovereign state, to suit our purposes. And so Telerica was born. It began as Teleria, then became Telephonia and Newphonia, but eventually G O hit on Telerica, which was easy to pronounce in various languages. Meanwhile, we'd come to the conclusion that advertising was not the best medium, and we decided to produce a booklet of digital case studies. When Richard was presented with this idea he actually showed signs of what, for him, goes for mild enthusiasm, and we were in business. A colleague and I worked on the case studies while Jeans (and the entire staff of his agency, to judge by the cost) created the Telerica environment.

It took a long time to produce the booklet, but it came out eventually and has proved valuable. We used the first chapter to describe the country—its geography, industry, population and tourist attractions, and its significant advances in education, industrialisation and telecommunications. The booklet appeared as a product of the Telerica Telephone Administration, with hardly a mention of Ericsson, and when we offered it in advertisements later, they were signed off by TTA, care of Ericsson.

Our next step was to use the Telerica theme for exhibitions as well, the Miami show of 1981 being the first occasion. G O and Richard pulled out all the stops, and fitted up the new country with a flag, a national anthem, a national drink, a national newspaper—and an ambassador to the United States. For this post, I was elected, and at the briefing for Ericsson staff I was presented with a scroll describing my position in the following words:

TELERICA
Be it known that our trusted and wellbeloved
John Meurling
being a steadfast and faithful servant of the
Sovereign State of Telerica is this
day appointed
Ambassador Plenipotenciary (Technical)
to the United States of America and to such
surrounding states as it may please us to direct him
And that
he shall in his appointed role represent the best
interests of the said sovereign state
And that
in pursuance of these interests he is hereby
empowered to make such Overtures Enoblements and
to carry out such Summary Executions as may be
deemed necessary

Witness our hand this 19th Day of March 1981
Signed: G O Douglas
Richard Jeans
Victor Henson
Bill Livingstone

"In colloquio Pax"

We've used Telerica as a theme on several occasions now. We've produced a second booklet of case studies, and we've found the material very useful for our training programs. The content is strictly technical/economic, but with the gimmicks and humour added we have a happy and effective combination.

CHAPTER 6: AS FAR AS WE CAN SEE

THE COMPETITION TODAY

TODAY, there are about a dozen competitive switching systems on the international market. In theory, at least, all of them possess the characteristics which are called for to build digital networks, though they are not all fully digital. They are at different stages of development; some of them are well-established, and proven in service, while several others are still unknown quantities.

Roughly speaking, the world market today absorbs switching equipment worth 20 to 25 billion dollars a year. This is largely made up of new exchanges to cater to the growth in the number of subscribers. The correlation between gross national product and number of telephones per capita is well known, and proves, of course, that with growing wealth, a country installs more telephones. (Or, if you prefer, you can see it as proof that telecommunications are a prerequisite for growth in GNP.) There is also now a second market, represented by the need to replace old switching equipment with new. There is still plenty of equipment in service today that was installed back in the '20s, over 50 years ago. The time has come to replace such equipment—and indeed, as digital switching and ISDN become necessities, we can also expect those exchanges which have been installed fairly recently to be phased out more rapidly.

Expressed as the number of exchange lines in public exchanges, the market amounts to some 30 million lines per year. A comparison of this figure with the number and resources of the manufacturers in the market today raises an obvious question: is there not, in effect, some rather alarming over-capacity in the industry? Now that the equipment is almost totally electronic, manufacture is based to a very high degree on automated production techniques, which call for long runs and high volumes. The industry is also increasingly politically influenced, with many countries set on establishing home manufacture either of a national design or under license. Local manufacture used to be desirable as a provider of jobs, but its role has changed. Today telecommunications represents high technology—computers, electronics, software—and is seen as important for developing other local industries, education and national know-how.

There has been much speculation on the future of the industry. And in the last few years, certainly, there has been much restructuring. These are harassing times—especially for our competitors.

Late in 1982, it was announced that the USA's AT&T was talking to the Netherlands' Philips. Exactly what the outcome of the talks will be is not yet clear, but public switching is involved and it looks as though AT&T, who have had very limited success in tackling markets outside the US on their own, may be trying a new approach. The only logical outcome would seem to be that Philips should take over the engineering of the No. 4 and No. 5 ESS systems for international markets—a major design effort—and that Philips should also add marketing muscle. If so, what is to happen to the present Philips market, where the PRX system is installed—Peru, for example, Indonesia, Holland and Saudi Arabia?

The case of ITT is even more interesting.

ITT is promoting System 12 heavily. System 12 is really a series of three different products, and of these it is the 1240 that is the most interesting and most heavily marketed at the moment. ITT has an impressive order backlog, but few installations which are commercially operational.

The ITT reference list shows high numbers. This is partly because ITT uses a different method of reporting, so that numbers on order include what we call long-term agreements for system choice, without the placing of firm legal contracts. ITT also uses a conversion factor—between 2 and 4, as far as we can tell—to translate transit lines into equivalent subscriber lines, while Ericsson does not apply a multiplication factor. These points may be arguable.

The 1210 was developed by, and is marketed by, ITT's US subsidiary, North Electric. (In fact, the development of 1210 began while North Electric was still subject to a certain amount of Ericsson influence.) System 1210 has been sold to the Philippines, Taiwan, Trinidad, and the Virgin Islands, plus at least half a million lines in the United States.

The second system in the System 12 family is System 1220, of which 6000 lines have been contracted for Italy.

System 1240 is the main product for the markets in which ITT and Ericsson compete. We have been given to understand that the first commercial exchanges will be in service during 1984.

Meanwhile, ITT has been doing a bit of regrouping. In 1982, it sold out its remaining French company, CGCT, to the French government. ITT had already pulled out of Brazil, and, also in 1982, it sold out its majority interest in the British subsidiary, Standard Telephones and Cables. We don't know all the reasons for these moves, but when we take into account the tenders that ITT has lost to Nippon Electric, CIT-Alcatel and Ericsson, we find that ITT is no longer quite the formidable competitor it used to be. We are certainly very curious to see how System 1240 turns out when it comes into operation.

We've also been watching Thomson-CSF with interest. After our initial collaboration over the AXE licence, which has led to some 900,000 lines installed or on order in France, our ties with the French have become thin. Thomson decided early on to develop its own digital switch, instead of extending the AXE licence, and the MT system was introduced to the market in the late '70s. Thomson now has a large backlog of orders from the French PTT, and between 1978 and 1981 managed to beat us and the rest of the competition through imaginative marketing to contracts for Greece, Colombia, Iraq and Chile, among other countries.

It seems that Thomson has had a series of problems getting the MT system airborne. Thomson now describes the development time as six years instead of the three originally planned, and the development costs as 1 billion francs rather than the 500 million which was the original estimate. In some markets, Thomson's contracts have been converted to CIT-Alcatel.

Let's look at Nippon Electric, and the NEAX 61 system. In all fairness, one has to concede that Nippon Electric has done a good job. The system was first presented in 1977. By 1979, a multi-processor configuration was on offer, and during the summer of 1982, Nippon Electric cut into service the so-called Cinturón project in Buenos Aires. This is a system of digital tandem exchanges which also incorporates 34 and 140 megabit fibre cables, and is undoubtedly a prime reference installation. If NEC now gets the very large local exchange in Syria going successfully it has certainly found a position among the three current leaders in switching.

The local exchanges for Buenos Aires are to come from Siemens. In 1974, with three trial installations in service with the Bundespost, Siemens had come a long way with the design of a new SPC system called EWS. Then, in 1977 I believe, the Bundespost got the digital message, and decided that its future system would be digital. Siemens had to go back to the drawing board. The original plan was to use the SP 103 control system from the analogue switch, but in 1981 Siemens announced a new control system (SP 113) based on an Intel microprocessor. There is now a situation with two

Siemens systems—EWSA (analogue) and EWSD (digital)—and some confusion, at least within Ericsson, as to which is going where. But we know that when the initial problems are over, Siemens will be a force to be reckoned with.

And Britain? Guided by what was then the British Post Office, the British telecommunications industry was quick to explore the potential of stored program control. A field trial in 1965 led to the development of two electronic systems, TXE2 and TXE4, both analogue. TXE4 in particular, which is manufactured today by STC, has been installed in large numbers in the UK, though very little has been exported.

Today, System X, which was largely designed by GEC and Plessey to British Telecom's specification and with British Telecom's financial support, is attracting attention from us and a lot of other people—it is now being marketed abroad, and the first export order has been booked.

At Ericsson, we are proud to have supplied a series of switching systems for London's international exchanges—crossbar in the early '70s, then a large AKE stored program control system, and now, in the spring of 1984, the first phase of the Keybridge installation with AXE.

CIT-Alcatel stands out as the first company to promote the concept of digital switching and to develop the first commercially viable product, the E10 system. We started to hear about the E10 back in the early '70s, when a lot of attention was focussed on the small town of Lannion in Brittany. Lannion was where the first E10, a central group switch and a number of remote units or concentrators, was put into service. Since then, the system has been sold in large volumes in many parts of the world. The original design, E10A, is a fairly small exchange which handles up to some 12,000 lines. The E10B is a development of it, and has been sold to many countries, including the Yemen, South Africa, Lebanon, Costa Rica and India. At present, the E10B incorporates an analogue subscriber stage, but a digital stage has been announced. CIT-Alcatel also has a transit switching system called the E12. E12 was installed in one exchange for the French PTT, but it was discontinued when the PTT planned to go over to the MT20. However, with the delays experienced over the MT20, the PTT reverted to ordering more E12, and CIT-Alcatel is also offering the system for export. CIT-Alcatel also has a fully digital system called the E10S, designed for rural applications.

American Telephone and Telegraph (AT&T), Bell Labs and Western Electric, now AT & T Technologies, the world's largest telecommunications producer, have nearly always been in the lead in the telecommunications industry. Happily, they have not been active in the export market since the 1920s. Until recently, we have only met them in a few carefully-selected markets, but as the break-up of the Bell empire starts to take effect, we shall certainly see a lot more of them. They tendered unsuccessfully for the first Saudi contract, and later on secured the first large local exchange contract from the Republic of Korea. We have already mentioned the Philips alliance, which will in time take them into the international market. The main products are the No. 4 ESS for transit exchanges, and the No. 5 ESS for local exchanges. Both were designed for the USA, and even if Philips is working on the necessary modifications the project is bound to take some time. It is difficult to envisage a system working in a non-US market and in volume production before 1986.

In sheer number of digital installed lines in service, Northern Telecom with its DMS system currently tops the league. With DMS 100, Northern Telecom has sucessfully penetrated both the independent companies and the Bell system in the US—no small achievement, even if there were some tough financial problems for a time. Northern is actively looking for more markets, outside the US, and has had some success. The modification of the DMS 100 for the CCITT/CEPT standard is, we understand, being handled by the Austrian company Kapsch/Schrack.

To complete the list of significant members of the switching system mafia, several more manufacturers should be mentioned. GTE of the USA not only supplies the US and Canada, but also has interests in Europe, and is casting its eyes over Latin America and the Far East. And in Japan, there are three suppliers apart from NEC: Fujitsu, Hitachi, and OKI. Fujitsu is having some export success, with contracts placed by Hong Kong, Singapore and Colombia. It is not clear what the Japanese manufacturers will do about the new system, called the D 60, designed by NTT, the Japanese Administration. Will it be offered for export, or will these companies be supplying different products to the Japanese and export markets?

In Finland, the Nokia company has begun to compete successfully, with digital switching systems based originally on the CIT-Alcatel design. In Brazil and Italy efforts to design a national system are under way. And so on. It's an open question whether there is a big enough market to support all the systems, and whether all these companies will survive. Some restructuring is inevitably taking place, and there are several examples of major companies moving into other products and other markets—notably data-processing and office automation.

The common denominator in all the problems is processing power and software. A real-time data-processing system to handle up to 500,000 busy hour call attempts is not easy to design. Several different solutions are being suggested, the two main ones being multi-processor organisation and, quite simply, faster units. For System 12, ITT has chosen distributed control. Distributed control offers attractive advantages on paper, but in practice may result in a very large number of processors.

Where, in all this, does AXE stand today?

In June 1984, as we were trying to complete this story, AXE had been sold to some 90 Administrations, in 52 countries. The number of lines in service was over 4 million, the number on order over 5 million.

AXE-based mobile telephone networks were in service in the Nordic countries, in Saudi Arabia, in the US, and in Spain. We had our second contract for Korea, and orders for large installations in Malaysia and Thailand. Both were very valuable to us, since though we had been supplying them for years, there had been signs that the Japanese were overtaking us.

We also had our first contract for the supply of AXE switches to the United States; there were orders for the People's Republic of China (Peking and the Canton region); and for Saudi Arabia, over a million lines were installed or on order, as additional contracts had been placed.

A large market for AXE, though not for Ericsson, was France, with about 900,000 lines on order or in service—that the number is so large may be attributed to the delays on Thomson's MT20/25 system.

The complete reference list is on pages 172 and 173.

The list includes systems for mobile telephony (cellular radio). Adding up the total we arrive at 9.7 million lines, of which over 4 million were in service. I should stress in service rather than just installed. And during July we continued to clock up orders and passed the 10 million line mark. This would have been a good occasion for another serving of Swedish pastry—but nobody took notice. We are pretty busy.

In Norway, we were beaten by ITT. The Norwegian Administration chose System 12, in preference to AXE and a number of other systems. This surprised us and a lot of other people. To lose a market so close to home was a bit of a come down. But subsequent successes in other markets have proved that AXE is still in the lead. Early in 1984, AXE was selected by the

JUNE 1984

	On order Multiple capacity			Installed Multiple capacity		
	No. of new exch.	Local	Transit	No. of exch.	Local	Transit
Antilles (Neth.)				1	10 000	
Argentina	3	17 000		4	25 000	
Australia	71	282 692	108 300	17	66 632	14 900
Bahrain		4 000		1	13 000	6 000
Brazil	22	185 440	1 674	13	105 600	4 000
Cayman Islands		256		1	3 456	450
Chile				1		4 000
China	3	40 000	3 500	1	7 500	
Colombia	19	195 000	40 444	24	153 000	50 892
Costa Rica				1		8 000
Cyprus	8	44 288				
Denmark	109	415 996	54 550	15	51 720	42 844
Egypt				2		22 528
El Salvador	5	40 576	1 900	4	14 336	5 800
Fiji	1	3 000				
Finland	21	112 650	46 080	13	87 132	26 624
France	3	26 624		66	871 075	
Hong Kong	1		6 144			
Iceland	3	6 444		2	6 500	
Ireland	9	163 944	18 180	9	52 224	34 300
Italy	8	81 920	10 560	14	49 024	25 920
Jamaica	1		1 534			
Kenya	1		1 000			
Rep. of Korea	79	755 384	69 632	11	256	171 008
Kuwait	14	207 360	23 552	7	120 696	13 312
Lebanon	2		8 192	1		6 656
Lesotho	5	6 656	4 096			
Macau		17 408		1	6 144	
Malawi	1	4 000				
Malaysia	42	460 720	19 144	9	122 880	5 120

JUNE 1984 *(Cont.)*

	On order Multiple capacity			Installed Multiple capacity		
	No. of new exch.	Local	Transit	No. of exch.	Local	Transit
Mexico	43	144 016	120 832	17	51 500	21 504
Morocco	9	68 800	3 500			
Netherlands	9	124 928	12 977	46	251 558	7 680
Norway	1	30 000		2	40 000	
Oman	1	4 500				
Pakistan	3	30 000	2 560			
Panama	2	8 992	460	6	28 000	1 300
Saudi Arabia	8	76 660	1 440	50	405 771	181 920
Seychelles				1	3 375	
Spain	15	82 960	47 616	16	99 920	44 032
Sweden	26	555 960	71 000	33	694 944	
Switzerland	2	8 700	12 288			
Thailand	29	295 296	2 050	1	5 120	
Trinidad and Tobago	1		2 048			
Tunisia	6	51 600		4	22 000	8 000
United Arab Emirates	2	15 488	3 210	9	67 840	18 434
United Kingdom	3	21 000	41 948	1		13 312
United States	3	15 000	19 200	3	1 300	19 200
Venezuela	39	259 256	8 200	3	23 000	
Yugoslavia	18	101 000	20 781	4	29 000	959
Zambia	1		1 000			
Zimbabwe	1		3 072			
Total	653	4 965 514	792 664	414	3 489 513	758 695

Swiss PTT, along with ITT's 1240 and Siemens' EWSD. This selection made us particularly proud, since Switzerland is a completely new market for us. In Switzerland, AXE is to be manufactured by the Hasler company.

We are going through hard times. There has been a world recession, and in most countries the economy is strained. Still, AXE has sold in numbers which in this business are quite exceptional. And though much of this story has concerned marketing, I am the first to concede that no amount of intelligent marketing, no number of elegant advertisements, no multi-colour brochures, and no professional presentations could have got these results if the product had not been outstanding.

The initial success of AXE was, of course, due to a very large extent to its arrival on the market at the right time. This, in turn, was the result of some careful planning, considerable strategic risk-taking, and an element of luck. AXE came on the market at a time when Administrations were ready and eager for a new technology. They had been through the necessary period of trial and evaluation; they had experienced the good and the bad; they were far enough up the learning curve to make the plunge into SPC. AXE, it is true, was new and unproven, but its design incorporated the experience gained with SPC of earlier generations.

The introduction of the digital version of AXE gave us a definitive lead over the competition—due, again, as much to timing as to the modernity of the product. Digital technology, with its dramatic effect on network costs, was beginning to be widely discussed within the industry and within such bodies as the CCITT and the CEPT. And at the time only three digital systems were on the market—AXE, Northern Telecom's DMS series, and CIT-Alcatel's E10.

It's been important for us to maintain that early lead—and we've done it, largely by planning. In the section called Growing Up, I've talked about the three continuous lines of development for AXE—new technology, new applications, and new functions. The whole system has been gradually developed to become the complete network node product for ISDN, the integrated services digital network. ISDN—the transformation of the digital telephone network into a public network carrying not only voice, but data, text and images as well—will be the next major development for many countries. AXE exchanges, present and future, will certainly be the heart of many of the ISDNs of the world.

If it sounds preposterous to claim that we still have a lead, a look at the world situation confirms it. If we glance at the competitive table we made in 1972 (page 45), the first thing that leaps out is that all the products listed (including our own ARE and AKE) have disappeared, replaced by new products. They're not dead, of course, they are still being sold and installed, usually as extensions of existing systems where an Administration is not yet ready to switch to something new. For any serious-minded Administration, switching switching systems is not something to be undertaken lightly.

In today's competitors' list, we find a whole new set of products, all digital. Their development began around 1976, and they have come on to the market since around 1980. Still, however, we can probably claim a two-year lead.

AXE is still the only system proven in major installations world-wide. Its performance everywhere, in all environments, and under every conceivable condition, has surpassed its specification, our expectation and our hopes. The Saudi project forced us through the industrialisation stage, a stage still to be gone through by some of our chief competitors. We describe AXE as future-proof—not only for our customers, but for Ericsson.

Part of our confidence is based on our conception of what the competition consists of. In our world, it is not the detail of a design, a new component, a new subscriber facility, or something similar, that defines how good a system is. It is system structure, or architecture as it is often called, that separates the men from the boys. And in structure, the software is decisive. AXE introduced a new way of thinking in software—a recognition that software is, so to speak, a living organism, not something stable. Even with the maximum care in design and testing, it will somehow, somewhere, have bugs in it. These bugs will show eventually, maybe after several years of operation, and must then be corrected. Software must also be modified and added to, to accommodate new functions and new applications, or to control new hardware. So the most important requirements set for our AXE system software were contained in words like flexibility, manageability, ease of handling, ease of modification, and so on. The result was AXE's functional modularity—a principle which has had world-wide acceptance.

So when we ask whether AXE can stand up against these new designs, initiated years later and to that extent more "modern", the important point, I believe, is that they are all "me too" systems. The basic features of their structure are no more modern than that of AXE. Will the new hardware be more competitive in capital cost? Will handling costs be lower? We must keep an open mind, but—I doubt it.

FINAL CHAPTER—BUT NOT THE END

BBC Television in the UK made a programme some years ago, as part of a series called "The Risk Business", looking at telecommunications as an export product. The programme seemed to be designed chiefly to knock British Telecom, but it also set out to show how the world's telecommunications companies—the dozen or so large firms that make up what I have called the international telephone mafia—are busily "carving up" the markets. As I hope the story of AXE has shown, "carving up" is definitely a misleading description of what goes on in this business. It is true that some markets are reserved for selected manufacturers, usually the national ones, and it is also true that some markets, for political reasons or by tradition, have kept to certain suppliers for a long time. The French, for example, are strong in the French-speaking countries of Africa. The British used to dominate the Commonwealth countries; the Swedes were strong, but far from alone, in the Scandinavian countries; and so on. For years, the US was supplied by American manufacturers only, Western Electric supplying the Bell operating companies, and GTE, Stromberg-Carlson, North Electric and a few other US manufacturers supplying the independent operators. (The situtation in the US is different today. Northern Telecom came in from Canada, and then Nippon Electric tried and now several European manufacturers, including Ericsson, plus ITT and Digital Switch, have announced their intentions to go for a piece of the divested AT&T cake.) Japan, of course, has always been a closed market, and there are no signs yet that this will change.

But the battle for markets has been a real one, with no cosy allotment of areas by agreement between multi-national organisations. The weapons in the battle are mostly technology, engineering intelligence, and marketing wit—plus hard work. To a very large extent, the troops have been engineers, whether they are working in engineering or in marketing. Also some of the generals, strategic wizards or tactical foxes, earned their scars and their early medals as engineers.

Of the two of us writing this book, one is an engineer, who has spent most of his time in marketing. The other is an advertising man and a professional observer of mankind. Before starting on the book, and while we have been writing it, we have spent many hours discussing the engineer, and we are both fascinated by him as a species of humanity. Engineers are not well documented in literature. A few of them have made their marks in history inventing the steam engine or building spectacular bridges. One of them invented the transistor, and that has certainly made a difference to a whole lot of us. But statistically, very few engineers actually *invent* anything, or design anything that earns a page in the annals of the human race, or the chronicles of technology. If we work in design, most of us will be found in groups of designers making up a team. If we work in marketing, we will be found putting together proposals, or looking after the logistics for getting equipment installed and maintained in service, or presenting and arguing and selling our wares. It is a not very exciting and pretty anonymous life, perhaps. But to many of us, it's a rewarding life, with enough satisfaction and some moments of joy.

Like me, many people at Ericsson have spent their whole professional lives in the company. With the years, we, too, become products of the company, for better or for worse. Like all Ericsson products, we develop and change; and like AXE, we represent the company, we are part of the Ericsson image, we are judged by our quality, our professionalism. The final judgement will be made by others, but I believe that the results show that the large team of people, mostly engineers, who built the AXE system did a hell of a fine job.

But reaching cruising altitude with AXE does not mean we can throttle back, relax and enjoy the view, as this condensed story might imply. It was, and it still is, immensely hard work. And in looking back, I have, no doubt, as is human, tended to gloss over many of the problems and difficulties, the set-backs, the nights and weekends spent in the office or the labs.

The spirit of adventure and challenge so evident at Ellemtel during the first half of the '70s has altered. The pioneering days are over, and bureaucracy has crept in. It is inevitable; the project is now too large to be managed by a band of individual enthusiasts. We need law and order, we need a systematic organisation—even though it often means that a small problem, which might once have been solved over a beer in the Ellemtel sauna, now takes a lot more time and a lot more paper. Some of the adventure has gone out of AXE, but AXE now confers status, and working with AXE is popular and satisfying. The original small group of people has grown to several hundreds, in Sweden and in many of the Ericsson companies outside Sweden. Often, these people are young, and relatively new in the Group. They grew up with electronics and computers at school and university, and they find it natural to be working on AXE. Other members of the team are old hands; we know a lot about telephony and cable networks and frequency division multiplexing, but we have had to go back to school in our old age and start learning new technology. To some of us, the experience has been painful, but we've mostly survived the great technological revolution, and the mix of the new bright lads and experienced old sweats is now a most valuable resource.

My American friends sometimes ask, "With all this hard work, and the spectacular success of AXE, what do you get out of it?" Well, Swedish practise is different from American practise. In Sweden, there are virtually no monetary perks, Swedish taxation would make them worthless. Yet Swedes work hard, most of them and most of the time. I suppose the challenge of the AXE programme and the excitement of taking AXE to market were a bonus for many of us. Most of us are not paid for overtime (again, the taxation would make that silly), but we devoted many nights and weekends to getting the job done. As we did so, our own capability grew, and some of us have been promoted.

Only a few people have entered this story by name, and I have refrained from trying to identify who was most important or who made the most useful contribution. But you may like to know what happened to some of us. Björn Svedberg became vice president and chief technical officer of the Ericsson Group in 1976, and the following year he was named president and chief executive officer—yet again, an engineer is head of the Ericsson Group. Way back in the beginning, Kurt Katzeff was chief technical person of the X Division, with a long background of distinguished work in system design. In 1972 he suddenly left, lured away by ITT, where he joined the System 12 team. A few years ago, he came back from Belgium, and is now with the Swedish Telecommunications Administration. Goran Hemdahl, who also joined ITT at one stage, now owns and runs his own consultancy. Three or four of the original design engineers have spent several years in Brazil, sponsored by the ITU. Claes Cederstrom, who was instrumental in getting together much of the material needed to get Södertälje into service (he was seen on several occasions transporting printed circuit boards by bicycle) later went to Thailand as head of switching systems marketing. He is now back at L M Ericsson, heading the department responsible for developing operation and maintenance systems. Torbjorn Johnsson, who headed up the early AXE proposal group with me, is now head of X Division's personnel department. Torbjorn Andersson, who headed Scandinavian marketing, is now chief of technical services in X, while Goran Sundelof, who for many years headed the first AXE design department at X Division, has taken over Scandinavian marketing. In 1979, Kjell Sandberg took over my old position in X Division. It is a staff function with the director of marketing, but it is very much a job that develops with the person in it.

Hakan Ledin was head of the X Division during the exciting and crucial years when AXE was taken into large scale production as a result of the Saudi Arabia contract, also the years when we were so busy participating in the many tender competitions. In 1979 he was elevated into top management and became instrumental in laying down, and implementing the strategic plans for guiding the Ericsson Group into data processing and office automation. Today he heads our activities in the US, including our efforts to introduce AXE, while retaining his position as a member of the executive committee of the Group.

Ove Ericsson was for many years head of the department marketing X Division products in Europe. He was one of the key figures in our getting the French contract, and later headed the Saudi Arabia effort—while he was still looking after Europe *and* playing golf. In 1979, he became head of X Division, and in our new organisation he is now head of the Public Telecommunications Business Area.

Hans Flinck was director of marketing for many years within X (and my boss). In early 1979, we both moved out and formed a small product and marketing planning group for the whole Ericsson Group. Hans continued to put a lot of time into X, while I was increasingly involved in trying to plan certain of our activities in the US. Hans moved back to a senior staff function with Ove Ericsson in 1981, and against all wishes retired in 1984.

Bengt Gunnar Magnusson was project leader and chief of the design team at Ellemetel during the critical first six years, and his contribution was enormous. He was singled out to receive the gold medal for engineering excellence of the Royal Swedish Academy of Engineering Sciences (and a couple of other similar decorations).

In 1976, Bengt Gunnar moved to a position as chief technical adviser at Ellemtel. In 1980, he transferred to Ericsson, and joined my little group, but only for a short time. He is now back at Ellemtel, heading an advanced projects department. He is still the whiz-kid of high-technology telecommunications, a great outdoors man, and only ever serious about technology and system design. I've seen his medals, but I've never seen him wear them. They would look silly with a sweater and jeans.

Kjell Sorme was one of the original members of Ellemtel's design team, first as Bengt Gunnar's number two, and from 1976 as head of the department. Then he too transferred back to Ericsson and eventually became head of the technical departments within X Division. And now he is making a transatlantic move, he has become a Texan and is heading up the Network Systems Division in the US; in American parlance network systems equals transmission systems, i.e. AXE. Hopefully another switch in time.

Most of us have lived to see and take part in the success of AXE. Sadly, Björn Lundvall and Fred Sundquist, both very much liked and respected, and both instrumental in bringing about that success, have died.

And Meurling? In 1981, I was still commuting to the west coast of the US, but my time was up. At the beginning of 1982, I was appointed head of the public relations department for the Ericsson Group, and to everybody's great surprise, I was elevated to the level of vice president.

I was no longer an engineer!

Public relations was something quite different from what I had been doing for most of my life. My engineering background and extensive knowledge of this big company were assets. But although (aided and abetted by Jeans) I had meddled at one time and another in composing advertisements of striking power and beauty, when I became ultimately responsible for their final appearance they didn't always come out right.

Another difficulty was simply keeping abreast of everything that was going on in this complex organisation. If I missed lunch with the management, I missed the news, and might end up reading all about it in the next day's newspapers.

In any case, journalists are a mixed blessing. They now figured largely in my life, and it took great self-discipline and a good deal of black humour not to land up in constant feuds with them, anyway some of them.

The job lasted less than two years, and then my life took yet another turn: I became a one-man splinter group, with the task of getting an investor relations programme going.

We had never entirely neglected our shareholders. We had sent them annual and quarterly reports, and a little periodical, but not much more. However, since about 1980, American investors had become increasingly interested in Ericsson shares, and in the Spring of 1983, we issued over 4 million new shares. Now, more than 30 percent of the company's shares are owned by Americans—mostly institutions. US investors and securities analysts are a different breed from their European counterparts. The US version is inquisitive, nervous, professional and vocal, and wants to see his investments grow. If they don't, he sells. And in the eyes of the US investor, our greatest fault was lack of information, coupled with the lack of someone to talk to without making international telephone calls.

So, for a start, I began to spend a lot of time in New York. I got myself an office and an apartment in the Big Apple and became a transatlantic commuter, with the usual strong symptoms of schizophrenia as a result. I'm doing it now (June 1984), and it's a fascinating and worthwhile experience. I've never in my life spent so much time on the telephone, answering, or trying to answer, every conceivable form of tricky question. I don't understand why the share price isn't soaring, instead of dropping as it has done lately.

I'm doing another job, too—in fact, I have two jobs and four bosses, which means *real* schizophrenia. In this second job, I'm back in public switching and engineering. Planning (my business card even says "strategic planning") is what I'm supposed to be doing, but the adrenalin comes from our new drive to market AXE in the US. The divestiture of AT&T has suddenly produced seven autonomous new operating companies—a very large, diverse new market replacing a virtual monopoly.

It is thrilling to be a member of the team now approaching this market. What the results will be is naturally an open question, but we're heavily and single-mindedly committed—including old Meurling.

And so let's leave him there, the teller of this engineer's tale, back in the fold with his fellow engineers.